Maximilian Kloess

Hybrid and electric cars - An energy economic analysis

Maximilian Kloess

Hybrid and electric cars - An energy economic analysis

Potentials to reduce energy consumption and greenhouse gas emissions in passenger car transport

Südwestdeutscher Verlag für Hochschulschriften

Impressum/Imprint (nur für Deutschland/only for Germany)
Bibliografische Information der Deutschen Nationalbibliothek: Die Deutsche Nationalbibliothek verzeichnet diese Publikation in der Deutschen Nationalbibliografie; detaillierte bibliografische Daten sind im Internet über http://dnb.d-nb.de abrufbar.
Alle in diesem Buch genannten Marken und Produktnamen unterliegen warenzeichen-, marken- oder patentrechtlichem Schutz bzw. sind Warenzeichen oder eingetragene Warenzeichen der jeweiligen Inhaber. Die Wiedergabe von Marken, Produktnamen, Gebrauchsnamen, Handelsnamen, Warenbezeichnungen u.s.w. in diesem Werk berechtigt auch ohne besondere Kennzeichnung nicht zu der Annahme, dass solche Namen im Sinne der Warenzeichen- und Markenschutzgesetzgebung als frei zu betrachten wären und daher von jedermann benutzt werden dürften.

Verlag: Südwestdeutscher Verlag für Hochschulschriften GmbH & Co. KG
Dudweiler Landstr. 99, 66123 Saarbrücken, Deutschland
Telefon +49 681 37 20 271-1, Telefax +49 681 37 20 271-0
Email: info@svh-verlag.de

Approved by: Wien, TU, Diss., 2011

Herstellung in Deutschland:
Schaltungsdienst Lange o.H.G., Berlin
Books on Demand GmbH, Norderstedt
Reha GmbH, Saarbrücken
Amazon Distribution GmbH, Leipzig
ISBN: 978-3-8381-2902-0

Imprint (only for USA, GB)
Bibliographic information published by the Deutsche Nationalbibliothek: The Deutsche Nationalbibliothek lists this publication in the Deutsche Nationalbibliografie; detailed bibliographic data are available in the Internet at http://dnb.d-nb.de.
Any brand names and product names mentioned in this book are subject to trademark, brand or patent protection and are trademarks or registered trademarks of their respective holders. The use of brand names, product names, common names, trade names, product descriptions etc. even without a particular marking in this works is in no way to be construed to mean that such names may be regarded as unrestricted in respect of trademark and brand protection legislation and could thus be used by anyone.

Publisher: Südwestdeutscher Verlag für Hochschulschriften GmbH & Co. KG
Dudweiler Landstr. 99, 66123 Saarbrücken, Germany
Phone +49 681 37 20 271-1, Fax +49 681 37 20 271-0
Email: info@svh-verlag.de

Printed in the U.S.A.
Printed in the U.K. by (see last page)
ISBN: 978-3-8381-2902-0

Copyright © 2011 by the author and Südwestdeutscher Verlag für Hochschulschriften GmbH & Co. KG and licensors
All rights reserved. Saarbrücken 2011

Abstract

Fossil fuel dependence and emissions of exhaust gases and greenhouse gases are some of the major problems passenger car transport is facing today. More efficient and cleaner propulsion technologies are one approach to alleviate these problems. The recent development of car powertrain electrification heads in this direction. Every year new hybrid models are presented and the introduction of plug-in hybrid (PHEV) and electric cars is scheduled for the upcoming years. There are high expectations that this beginning process of vehicle powertrain electrification could lead to pure electric passenger car transport one day. However, the key factors that influence this development as well as the possible time horizon and the potential effects on the energy demand are to a high extent uncertain today. This thesis tries to shed light on some of these uncertainties. The thesis tries to give answers to the following questions:

o   What is the economic performance of electrified propulsion technologies today and what are their perspectives for the future?
o   What are the crucial factors for the spread of hybrid and electric cars?
o   Within what time frames hybrid and electric cars can attain considerable market shares?
o   What role can policy play to encourage the spread of these cars and to improve the efficiency of the sector as a whole?
o   What are the main drivers of energy consumption and greenhouse gas emissions in passenger car transport?
o   How will large scale introduction of hybrid and electric cars affect the primary and final energy demand of the car fleet?
o   What is their potential to reduce GHG emissions within the transport sector?

The questions are approached systematically from an energy economic perspective: First a detailed techno-economic assessment of propulsion technologies of passenger cars is performed. Thereby, the entire range from conventional propulsion technologies to pure battery electric cars is analyzed. This includes hybrid electric vehicles (HEV) with different extents of hybridization (micro-, mild- & full hybrid), Plug-In hybrids (PHEV), battery-electric vehicles (BEV), as well as fuel cell vehicles (FCV). The results of the techno economic assessment show that battery costs and fuel prices are the key factors that affect the economic competitiveness of hybrid and electric cars. While hybrid cars are close to becoming cost effective, pure electric propulsion systems (PHEVs & BEVs) require a considerable reduction in battery cost and higher fuel prices. The results of the cost estimation 2010-

2050 indicate that hybrid systems will be the least cost option in a short term (up to 2020). With a reduction of battery costs and increasing fuel prices electric propulsion systems (PHEVs and BEVs) become the best mid- to long term option (after 2020).

To estimate the diffusion of electric cars and their effects on energy demand and greenhouse gas emissions of the passenger car fleet, a model-based analysis is performed. The applied model combines a bottom-up model of the Austrian passenger car fleet with top down approaches to model shifts in passenger car transport demand and transport service level. With the model scenarios for the time frame 2010-2050 under different political and economic framework conditions are developed.

Four main scenarios are analyzed that combine moderate and high fossil fuel price increases, with high and low degrees of regulatory policy intervention in the passenger car fleet. The results show that energy demand and greenhouse gas emissions cannot be reduced significantly by simply switching to more efficient hybrid cars. Considerable reductions require a deceleration in growth of transport volume and a true leap in efficiency of applied technologies. The scenario results indicate that fiscal measures and higher fossil fuel prices are the main catalysts for such a development. Higher taxes on fuels and on low efficient cars are driving a higher market share of electrified cars sooner. These measures also lead to lower average curb weights and lesser engine power of cars sold, a generally smaller fleet and lower yearly odometer readings of cars. All these effects cause a considerable reduction in energy demand of the fleet and an increasing importance of electricity as fuel. With a pure renewable electricity supply fossil energy demand of passenger car transport can be reduced by up to 86 % and greenhouse gas emissions by up to 68 % by 2050. The fact that the resulting electricity demand could be covered with domestic renewable electricity potentials points out the high potential of electricity as an energy carrier for road transport with respect to decarbonisation and diversification of the energy supply.

Kurzfassung

Die Abhängigkeit von fossilen Energieträgern sowie Abgas- und Treibhausgasemissionen sind einige der zentralen Probleme mit denen der Straßenverkehr heute konfrontiert ist. Effizientere und abgasärmere Antriebstechnologien sind ein Ansatz um diesen Problemen zu begegnen. Der jüngste Trend zur Elektrifizierung des Antriebsstrangs ist ein deutlicher Impuls in diese Richtung. Jedes Jahr kommen neue Hybridmodelle auf den Markt und auch die Markteinführung von Elektrofahrzeugen und Plug-In Hybriden ist für die kommenden Jahre geplant. Dies weckt die Erwartungen, dass dieser Trend eines Tages zum rein elektrischen Straßenverkehr führen könnte. Die kritischen Einflussfaktoren auf diese Entwicklung, deren Zeithorizont sowie deren mögliche Auswirkungen sind heute jedoch noch ungewiss. Ziel dieser Arbeit ist die Antworten auf folgende Fragen zu liefern:

o Wie ist die wirtschaftliche Konkurrenzfähigkeit elektrifizierter Antriebstechnologien heute, wie sind ihre Perspektiven für die Zukunft?

o Was sind die zentralen Einflussfaktoren auf die Wirtschaftlichkeit von Hybrid- und Elektrofahrzeugen?

o In welchem Zeitraum können Hybrid- und Elektrofahrzeuge nennenswerte Marktanteile erlangen?

o Welche Rolle spielen politische Rahmenbedingungen für die Verbreitung dieser Fahrzeuge und für die Effizienz des gesamten Sektors?

o Was sind die wichtigsten Treiber von Energieverbrauch und Treibhausgasemissionen im PKW Verkehr?

o Wie wird sich die Einführung von Hybrid- und Elektrofahrzeugen in großem Maßstab auf Primär- und Endenergieverbrauch der Flotte auswirken?

o Welche Potentiale zur Reduktion von Treibhausgasemissionen ergeben sich?

Zur Beantwortung dieser Fragen wird ein systematischer, energiewirtschaftlicher Ansatz verfolgt: Zuerst wird eine detaillierte techno-ökonomische Bewertung verschiedener Antriebstechnologien durchgeführt. Hierbei wird das gesamt Spektrum vom konventionellen bis hin zu rein elektrischen Fahrzeugen untersucht. Dieses umfasst Hybridfahrzeuge mit unterschiedlichem Grad der Elektrifizierung (Mikro-, Mild- und Voll-Hybride), Plug-In Hybride, batterie-elektrische Fahrzeuge sowie Brennstoffzellenfahrzeuge. Die Ergebnisse der techno-ökonomischen Bewertung zeigen, dass Batteriekosten und Kraftstoffpreise die entscheidenden Faktoren für die Wirtschaftlichkeit von Hybrid- und Elektrofahrzeugen sind. Währen Hybridfahrzeuge bereits heute annähernd konkurrenzfähig sind, erfordern elektrische Antriebsysteme (Plug-In Hybride und E-Fahrzeuge) eine Reduktion der Batteriekosten sowie höhere Kraftstoffpreise, um sich am Markt zu behaupten. Die Ergebnisse der Kostenabschätzung 2010-2050 zeigen, dass Hybridantriebe in den nächsten 10 Jahren (bis ca. 2020) die wirtschaftlichste Option darstellen werden. Mit der Reduktion der spezifischen Batteriekosten und steigenden Kraftstoffpreisen werden elektrische Antriebsysteme mittel- bis langfristig zu den wirtschaftlichsten Antriebstechnologien (nach 2020).

Um die Verbreitung elektrische Antriebsysteme und deren Auswirkungen auf den Energieverbrauch und die Treibhausgasemissionen der PKW Flotte zu untersuchen wird eine modell-basierte Analyse durchgeführt. Das eingesetzte Modell kombiniert ein *Bottom-up* Modell der Österreichischen PKW-Flotte mit *Top-down* Ansätzen anhand derer Veränderungen in der Nachfrage nach der Energiedienstleistung PKW-Transport, sowie im Niveau deren Erbringung *(Service Level)* modelliert werden. Mit dem Modell werden Szenarien unterschiedlicher politischer und wirtschaftlicher Rahmenbedingung für den Zeitraum 2010-2050 entwickelt.

Vier Hauptszenarien werden analysiert, welche moderate und starke Anstiege der Energiepreise mit niedrigen und hohen Grad politischer Einflussnahme im PKW Bereich kombinieren. Die Ergebnisse zeigen, dass der Wechsel zu effizienteren Hybridfahrzeugen allein keine deutliche Reduktion von Energieverbrauch und Treibhausgasemissionen bewirkt. Eine solche kann nur durch eine Verlangsamung im Wachstum des Transportaufkommens und einem deutlichen Sprung in der Effizienz der Antriebsysteme erreicht werden. Die Szenario-Ergebnisse zeigen, dass fiskalische Maßnahmen und höhere Preise fossiler Energieträger wesentlichen Treiber für diese Entwicklung sind. Höher Steuern auf Kraftstoffe und ineffiziente Fahrzeuge beschleunigen die Verbreitung elektrischer Antriebsysteme und führen darüber hinaus zu einer Verlangsamung des Flottenwachstums, zu kleineren und leichteren Fahrzeugen in der Flotte, sowie zu einer Reduktion der jährlichen Fahrleistung. All diese Effekte bewirken eine signifikante Reduktion des Energieverbrauchs der Fahrzeugflotte und einer steigenden Bedeutung von Strom als Energieträger. Mit Strom aus erneuerbarer Erzeugung kann deren Verbrauch an fossilen Energieträgern bis 2050 um bis zu 86 % und deren Treibhausgasemissionen um bis zu 68 % reduziert werden. Die Tatsache, dass der resultierende Strombedarf durch inländische Potentiale gedeckt werden könnte zeigt welches Potenzial Strom als Energieträger für den PKW Verkehr hinsichtlich Dekarbonisierung und Diversifizierung der Energieversorgung besitzt.

Executive Summary

Motivation
Increasing fossil fuel prices and greenhouse gas reduction commitments will be serious challenges for passenger car transport in the next years. More efficient propulsion technologies and low carbon fuels can contribute to the solution of these problems. Today electrification/hybridization of propulsion systems is seen as an appropriate measure to improve the efficiency of passenger cars. However, hybrid electric vehicles (HEV) are only the first step in a development that can ultimately lead to pure electric propulsion systems. With its superior efficiency and zero direct emissions battery electric cars or fuel cell cars are promising technologies for long-term improvement of efficiency in passenger car transport. Today, these vehicles are still facing serious technical, economical and infrastructural barriers. If they manage to overcome these they have high potential to reduce energy consumption and emissions of passenger car transport and they will fundamentally changes its energy supply.

Structure
The global objective of this thesis is to analyze how hybrid and electric propulsion technologies can contribute to the reduction of energy consumption and greenhouse gas emissions of the passenger car fleet on the particular example of Austria.

In pursuit of this global objective, the thesis addresses the following questions:

- What is the economic performance of electrified propulsion technologies today and what are their perspectives for the future?
- What are the crucial factors for the spread of hybrid and electric cars?
- Within what time frames can hybrid and electric cars attain considerable market shares?
- What role can policy play to encourage the spread of these cars and to improve the efficiency of the sector as a whole?
- What are the main drivers of energy consumption and greenhouse gas emissions in passenger car transport?
- How will large scale introduction of hybrid and electric vehicles influence the primary and final energy consumption of the car fleet?
- What is their potential to reduce GHG emissions within the transport sector?

To answer these questions a two-step approach is followed. First a detailed techno- economic assessment of hybrid and electric propulsion technologies is performed to analyse their economic competitiveness today and to identify key factors for their future potential. Based on these results a model based analysis is performed using a scenario model of the Austrian passenger car fleet. With this model market and fleet penetration scenarios are developed for the time frame 2010-2050 with different political and economic framework conditions.

Techno-Economic Assessment

Technically electrification/hybridisation of the powertrain is an effective measure to cut energy consumption and greenhouse gas emissions of passenger cars. Hybridisation can alleviate some of the main technical deficits of conventional propulsion systems that can be traced back to the specific operation characteristics of the internal combustion engines. In a hybrid system the electric drive supports the engine in order to run in its optimal operation point and recuperates breaking energy. The higher the drivetrain is electrified the better it can support the engine and the more efficient becomes the car. However, higher electrification also means higher complexity of the drivetrain, more powerful electric machines and higher battery capacity, altogether leading to higher system costs.

In the analysis various types of hybrid systems with different extends of electrification are considered including different types of hybrid electric cars (HEV), Plug-In hybrid cars (PHEV), battery electric cars (BEV) and also fuel cell vehicles (FCV). (see Figure 1)

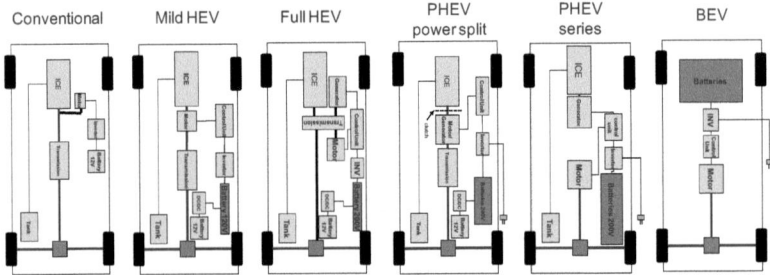

Figure 1: Conventional, Hybrid and Electric Powertrain Systems

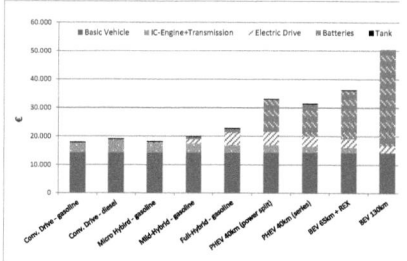
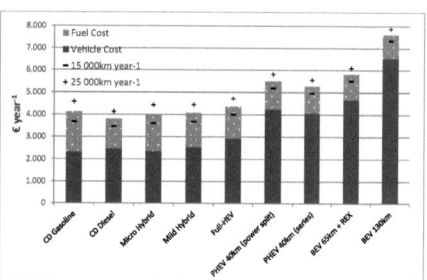

Figure 2: Net Investment Cost of Powertrain Systems in 2010 (Middle Class)

Figure 3: Yearly costs at 20 000 km year^{-1} – status 2010 (net vehicle cost & gross fuel price)

From an economic perspective the optimal degree of electrification is always a trade-off between system costs and fuel costs. The results show that with today's (2010) costs of electric components the gasoline price has to be at least 1.5 € liter^{-1} for hybrid systems to become cost effective. Below this price level only micro and mild hybrid systems can compete with conventional technology at average annual driving distances (15 000 – 20 000 km). Fully electrified propulsion technologies like plug-in hybrid electric vehicles (PHEV) need gasoline prices higher than 2.5 € liter^{-1} to compete. This points out that the costs of pure electric propulsion systems are still too high with batteries being the main cost drivers. In order to become economically competitive with conventional cars they will rely on a reduction of battery costs and increasing gasoline and diesel prices. However, the results show that even with considerable cost reductions batteries remain a cost driver making cars with long electric ranges economically unfeasible. Therefore, PHEVs with shorter electric driving ranges and an internal combustion engine as range extender have a better chance to address the mass market than pure battery electric cars in a mid-term.

The results of the cost estimation for the time frame 2010-2050 indicate that hybrid systems will be the least cost option in a short term (up to 2020). With a reduction of battery costs and increasing fuel prices Plug-In hybrid systems become the best mid- to long term option for middle class cars (see figure 4). At this condition battery electric vehicle (BEV) will become the first choice for compact class cars whose typical field of application requires lower driving ranges (e.g. urban areas). For both PHEVs and BEVs the economically optimal electric driving range will depend on the specific framework conditions (fuel price & yearly driving range) and the cost of batteries.

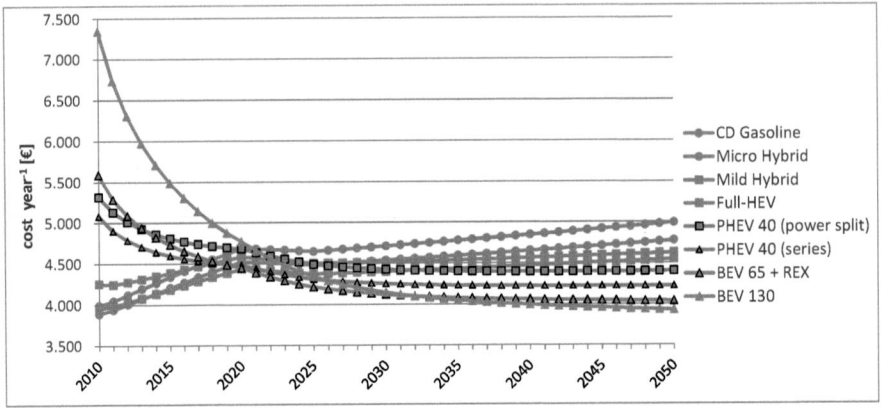

Figure 4: Estimated development of yearly costs of propulsion systems in the middle class 2010 – 2050 in the "High-Price-Scenario"

The assessment of hydrogen fuel cell propulsion systems has shown that fuel cell cost have to come down to 100-200 € kW^{-1} in order to economically compete with conventional technologies. It will be difficult to achieve the required reduction of fuel cell systems cost in a short- to mid-term, especially when considering that there is no bridging technology that could act as a driver for technology development. Unlike battery electric systems which can rely on hybrid technology to help reduce cost of batteries by driving their global cumulative production and generate technology spill-overs, there is no such technology link for mobile fuel cell systems. On the other hand fuel cell systems could solve two major problems of electric propulsion technology: storing enough energy on board for long diving distances and permitting fast refuelling. As long as these problems cannot be solved with battery systems, hydrogen fuel cells will remain in play as a long-term option.

Model based analysis

The model combines bottom-up and top-down modelling approaches and has been developed with the special focus on the analysis of effects of new technologies, fossil fuel prices and policy measures on energy consumption and greenhouse gas emissions in the Austrian passenger car fleet. The model captures the most important factors that affect energy demand of passenger car transport, like fleet growth, characteristics of new cars (mass, engine power, fuel consumption) and use of cars. The time frame 2010-2050 allows to analyse long-term effects of changes in economic and political framework conditions in the fleet. This permits the simulation of policy effects in a wider time horizon, which is especially relevant when long term carbon mitigation goals have to be met.

The model mainly consists of four modules (see figure 5):

Module 1: The first module is the vehicle technology model where the vehicle powertrain options are modelled bottom-up to analyse the influence of technological progress on their costs.

Module 2: The second module derives market shares of technologies based on their specific service costs considering different levels of willingness-to-pay. The heterogeneity in consumer preferences is modelled using a logit-model approach with specific service costs as the main parameter. The technology-specific diffusion barriers that arise from limitations in performance characteristics or lack of availability etc. are modelled by predefined constraints of maximal growth in market share of each technology.

Module 3: The third module includes the top down models that capture the influence of income, fuel prices and fixed cost on the demand for passenger car transport and transport service level.

Module 4: The fourth module is a bottom-up model of the Austrian passenger car fleet. The fleet is modelled in detail considering age structure, user categories and main specifications of the cars (e.g. engine power, curb weight, propulsion technology, specific fuel consumption, greenhouse gas emissions etc.). The settings are based on a data pool including detailed information about the fleet today and time series of historic developments between 1980 and 2008.

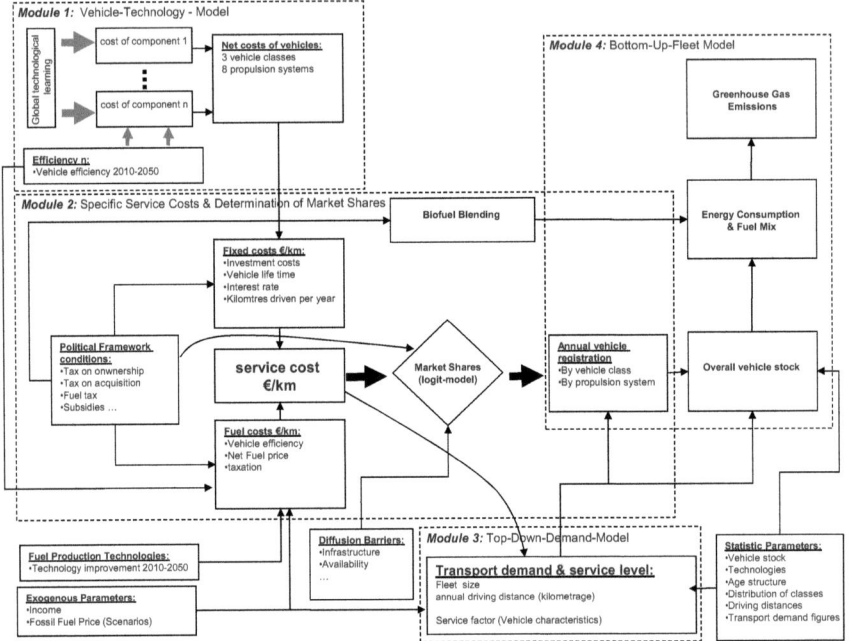

Figure 5: Scheme of the model

With the model four main scenarios are developed with two fossil fuel price scenarios and two policy schemes:

- "Business as usual"-Policy & moderate fossil fuel price increase (*BAU & Low Price - Scenario*)
- "Active" Policy & moderate fossil fuel price increase (*Policy & Low Price - Scenario*)
- "Business as usual"-Policy & substantial fossil fuel price increase (*BAU & High Price - Scenario*)
- "Active" Policy & substantial fossil fuel price increase (*Policy & High Price - Scenario*)

The **Policy** scenarios implicate major changes to the political and regulatory framework. Taxes are adapted with a clear focus on increasing energy efficiency and reducing greenhouse gas emissions of the sector (higher fuel taxes & higher tax on acquisition for cars with low efficiency).

The developed scenarios point out the key role of policy measures in passenger car transport. In the **BAU** scenario, where no major policy measures are taken WTW energy consumption and greenhouse gas emissions of the car fleet keep growing up to 2030 (WTW-energy consumption: +20 %; WTW-GHG emissions:

+14 %). This development is mainly driven by the growth of the car fleet (+27 % up to 2030), a relatively high yearly kilometrage and a high service level of cars. The diffusion of more efficient hybrid cars cannot offset the effects of theses drivers in the BAU scenario. Highly efficient fully electric cars (PHEVs & BEVs) only slowly diffuse into the fleet (only 12 % in 2030) and therefore show little effect on energy consumption and greenhouse gas emissions.

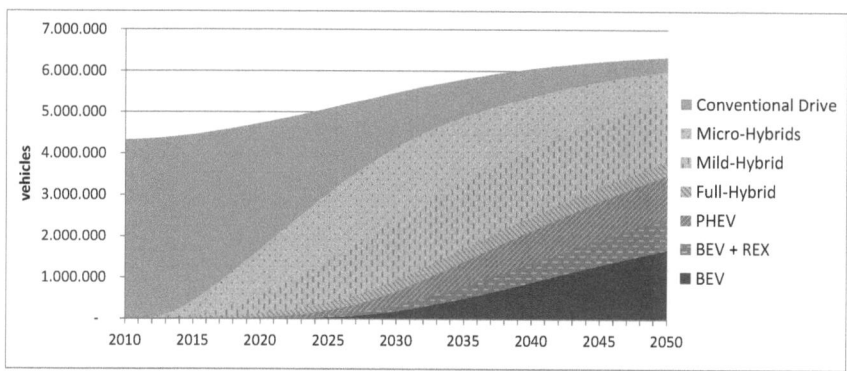

Figure 6: Development of the passenger car fleet: *BAU & Low Price – Scenario*

In the **Policy** scenario higher fuel taxes and higher taxes on inefficient cars lead to a significant reduction of both greenhouse gas emissions and energy demand in the fleet (fossil WTW-energy consumption: -30 % & WTW-GHG emissions: -26 % up to 2030). The higher fuel prices lead to a deceleration of the fleet growth (+9 % up to 2030), lower average weight and power of cars, lower yearly kilometrage and above all a strong diffusion of electric propulsion systems (36 % of the fleet in 2030). The latter is also driven by the vehicle taxes which promote efficient cars.

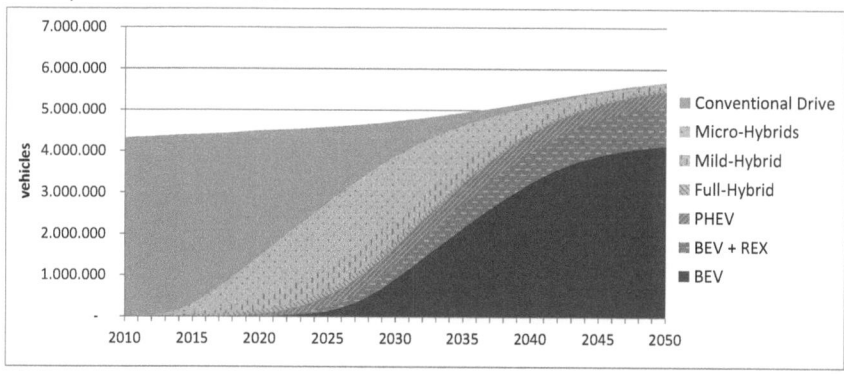

Figure 7: Development of the passenger car fleet: *Policy & Low Price – Scenario*

The comparison of the **Low Price** and the **High Price** scenarios indicate that higher fossil fuel prices also reduce energy consumption and greenhouse gas emissions by slowing down the fleet growth and fostering the spread of efficient propulsion technologies. However, the scenario comparison shows that the policy framework has considerably stronger effects.

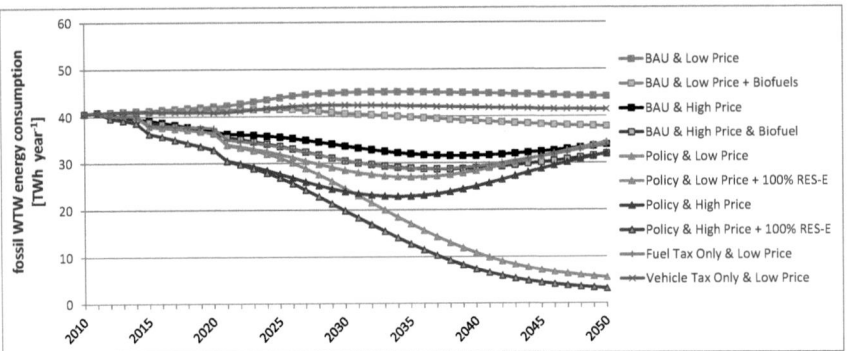

Figure 8: Fossil WTW energy demand of the passenger car fleet in the analyzed scenarios

The comparison of the electricity supply scenarios indicates that the full potential of greenhouse gas reduction of electric cars can only be exploited with a low carbon electricity supply. The **100% RES-E** supply scenario shows that a completely decarbonised electricity mix reduces the annual fossil fuel energy demand of the passenger car fleet by more than 86 % and greenhouse gas emissions by 68 % up to 2050.

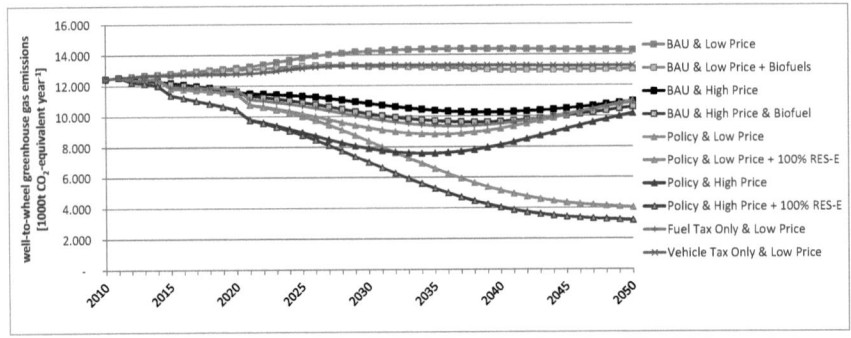

Figure 9: WTW greenhouse gas emissions of the passenger car fleet in the analyzed scenarios

All scenarios share one major trend: a shift in the passenger car fleet towards hybrid cars. Electrification can be considered a robust trend in automotive propulsion technology in the coming years and decades. This development will be mainly driven by increasing fossil fuel prices causing higher demand for efficient cars. Even though hybridisation will improve the efficiency of the fleet, the results of the *BAU-Scenario* point out that energy demand and greenhouse gas emissions cannot be reduced by simply switching to hybrid technology. They are not able to compensate the increasing energy demand caused by the steady growth in passenger car transport volume. Considerable reductions in energy consumption and GHG emissions require a deceleration in growth of transport volume and a more radical change in the applied technologies. Electric cars could offer the required leap in efficiency. They are three times more efficient than conventional cars and electricity as energy carrier can facilitates the decarbonisation of passenger car transport.

The results of the *Policy Scenario* indicate that a reduction of GHG emissions and fossil fuel dependence of the passenger car fleet can only be achieved by a combination of higher efficiency of cars, lower growth in demand for passenger car transport and a lower average service level of cars. Policy measures and new more efficient technologies are the main catalysts for this development. In the *Policy-Scenario* policies are set to support efficient cars by adopting higher taxes on fuels and on low efficiency cars, driving a higher market share of electrified cars sooner. These measures also lead to lower average curb weights and lesser engine power of cars sold, a generally smaller fleet and lower yearly odometer readings of each car. All these effects cause a considerable reduction in energy demand of the fleet and an increasing importance of electricity within the energy carrier mix. This then requires a low carbon electricity generation base in order to reduce overall GHG emissions.

Conclusions

The major conclusions of the thesis are:

- Hybrid cars are about to become costs effective today and their economic attractiveness will improve with increasing fuel prices in the future. This will lead to higher shares of hybrid cars in the market and in the fleet.
- Electric cars require a significant reduction in costs of key components (above all batteries) and higher fossil fuel prices to become economic attractive. Apart from the cost barriers, the technology has to face some serious acceptance barriers associated with its limited driving range, long charging time and infrastructure availability making the time horizon of large scale market introduction uncertain.
- Only fully electric propulsion technologies can deliver the required leap in efficiency, in order to bring about a substantial reduction in energy demand and greenhouse gas emissions in the next decades.

- Policy can push market and fleet diffusion of electric powertrain technologies by setting appropriate policy measures that support them. Higher taxes on fossil fuels and inefficient cars are effective measures to promote these efficient propulsion technologies. Furthermore, they can help to slow down the growth in demand for passenger car transport and lead to lower service level of the cars. These effects together can lead to a considerable reduction in energy consumption and greenhouse gas emissions of the fleet.
- Electricity as energy carriers for passenger car transport could significantly reduce the oil dependence of the transport sector. It offers a better diversification of the energy supply and a higher potential of decarbonisation than conventional fuels. In order to exploit the full potential of electricity as an energy carrier a low carbon electricity mix has to be aimed for.

Content

1 Introduction 1
   1.1   Motivation 1
   1.2   Objective 3
   1.3   Method of approach 4
   1.4   Main literature 5
   1.5   Structure of the Thesis 6
2 Energy economic parameters of passenger car transport in Austria 8
   2.1   Energy Demand of passenger cars in Austria 8
   2.2   Emissions of passenger cars in Austria 9
   2.3   The Austrian passenger car fleet 10
   2.4   Trends in passenger car sales 11
      2.4.1   Propulsion technologies 12
      2.4.2   Curb weight and engine power 12
      2.4.3   Fuel consumption and emissions 14
   2.5   User pattern of passenger cars in Austria 15
   2.6   Policy in the passenger car sector 17
      2.6.1   Tax on Acquisition/Tax on Registration: 18
      2.6.2   Tax on Ownership: 20
      2.6.3   Fuel Tax: 20
      2.6.4   Emission standards 21
3 Some theoretical background of energy economics and energy modelling in the transport sector 22
   3.1   Transport as an Energy Service 22
   3.2   Demand for Transport 23
   3.3   Energy demand 24
   3.4   Car Ownership 24
   3.5   Car use 27
   3.6   Service Level in passenger car transport 27
   3.7   The Rebound Effect in passenger car transport 28
   3.8   Costs of Transport 30
      3.8.1   Internal Cost of passenger car transport 31
      3.8.2   Specific service costs 32
   3.9   Transport demand and service level in passenger car transport in Austria ... 33
4 Technological change in propulsion technologies for passenger cars – A historical survey 36
   4.1   Technological Life Cycles and Technological Diffusion 37
   4.2   Diffusion of technologies for passenger cars 38
   4.3   Technological Learning 39
   4.4   Diffusion Barriers 41
      4.4.1   Lock-In phenomenon 42
   4.5   Historic technological trends in passenger car propulsion systems 43
      4.5.1   The breakthrough of the internal combustion engine 43
      4.5.2   Electric Mobility Hype (1990) 44
      4.5.3   Hydrogen & Fuel Cell Hype (2000) 45

- 4.5.4 Electric Mobility Hype (2010) 46
- 4.5.5 Key findings from past technology "hypes" 47
- 4.6 Stakeholders for the diffusion of vehicle propulsion technologies 48
  - 4.6.1 Consumers 48
  - 4.6.2 Car Industry 50
  - 4.6.3 Oil Industry 51
  - 4.6.4 Policy 52
- 5 Techno-economic assessment of hybrid and electric propulsion technologies for passenger cars 53
  - 5.1 Reducing Fuel consumption of motor vehicles 53
    - 5.1.1 Reduction of rolling resistance: 53
    - 5.1.2 Reduction of Aerodynamic drag: 53
    - 5.1.3 Reduction of vehicle mass: 54
    - 5.1.4 Improvement of engine efficiency 54
    - 5.1.5 Drivetrain improvements 55
    - 5.1.6 Electric Auxiliaries 55
  - 5.2 Electrification of the vehicle powertrain 56
    - 5.2.1 Conventional Drive (CD): 56
    - 5.2.2 Micro Hybrid: 57
    - 5.2.3 Mild Hybrid 57
    - 5.2.4 Full Hybrid: 58
    - 5.2.5 Plug-In Hybrid (PHEV) – parallel & power split drive: 59
    - 5.2.6 Plug-In Hybrid (PHEV) – Series Drive: 60
    - 5.2.7 Battery Electric Vehicle (BEV): 61
    - 5.2.8 Fuel Cell Electric Vehicle (FCV): 62
  - 5.3 Efficiency of automotive propulsion systems 63
    - 5.3.1 Vehicle Efficiencies in the analysis 67
  - 5.4 Electricity storage systems for Electric Cars 68
    - 5.4.1 Relevant Parameters for the assessment of batteries 69
    - 5.4.2 Battery Technologies 71
    - 5.4.3 Technological Progress of Batteries 74
    - 5.4.4 Specific costs of batteries 77
    - 5.4.5 Range Extenders 78
    - 5.4.6 Fuel Cells 80
    - 5.4.7 Electric range and curb weight in electric propulsion systems 81
  - 5.5 Economic Assessment of Electrified Propulsion Systems 2010 83
    - 5.5.1 Comparing Propulsion Technologies – Reference Vehicles 84
    - 5.5.2 Investment Costs 2010 86
    - 5.5.3 Fuel Costs 92
    - 5.5.4 Cost comparison of propulsion technologies 92
    - 5.5.5 Sensitivity Analysis 95
    - 5.5.6 Economic assessment of fuel cell systems 97
    - 5.5.7 Net present value of fuel cost savings through hybridisation 100
  - 5.6 Technological Learning effects of key components 102
    - 5.6.1 Battery Learning Curves 102

    5.6.2   Ressource Costraints of key materials 105
    5.6.3   Fuel Cell Learning Curves 106
    5.6.4   Fuel price and Fuel Price Scenarios 109
  5.7   Cost Scenarios 2010 – 2050 110
  5.8   Total cost in Austria 2010 112
6   Simulation model of the passenger car sector in Austria (ELEK-TRA-Model) 115
  6.1   Methodology 115
  6.2   Structure of the model 117
  6.3   Market shares of technologies 118
    6.3.1   Diffusion Barriers 120
  6.4   Modeling the demand and service level of passenger car transport 122
    6.4.1   Fleet development 123
    6.4.2   Modeling of the car user behavior 124
    6.4.3   Transport service level – shares of vehicle classes 125
  6.5   Bottom-Up Fleet Model 127
    6.5.1   Modelling the exchange rate of cars 128
  6.6   Energy Consumption and greenhouse gas emissions of the passenger car fleet   130
7   Assumptions for Scenario Development 132
  7.1   Global fossil fuel price levels 132
  7.2   Political Framework Conditions in Austria 2010-2050 134
    7.2.1   Business as usual (BAU) Scenario 135
    7.2.2   Policy scenario 135
  7.3   Specific Service Costs 2010-2050 136
  7.4   Market- and Fleet-Penetration 2010-2050 142
    7.4.1   Business as usual policy & low fuel price scenario (BAU & Low Price) 142
    7.4.2   Active policy & low fuel price scenario (Policy & Low Price) 143
    7.4.3   Business as usual policy & high fossil fuel price scenario (BAU & High Price)   145
    7.4.4   Active policy & high fossil fuel price scenario (Policy & High Price) 145
    7.4.5   Impact of technological learning effects 146
    7.4.6   Average characteristics of passenger car sales 147
8   Energy consumption, energy carriers and greenhouse gas emissions of the passenger car fleet in Austria 2010-2050 149
  8.1   Final energy demand in the scenarios (TTW) 149
  8.2   Fuels and fuel sources 2010-2050 151
    8.2.1   Energy- and greenhouse gas balances of cars and fuels 153
    8.2.2   Scenarios for biofuel blending 158
  8.3   Electricity as fuel for passenger cars 158
    8.3.1   Effects of Electric Vehicle Charging on the load profile 159
    8.3.2   Electricity supply scenarios 162
    8.3.3   Required electricity in the scenarios 164
  8.4   Well-to-Wheel (WTW) energy demand and greenhouse gas emissions 166
    8.4.1   BAU scenario 166
    8.4.2   Policy Scenario 167

		8.4.3 Scenario comparison	169
9		Conclusions	172
	9.1	Techno-Economic Assessment	172
	9.2	Model based analysis	173
	9.3	Outlook	175
10		References	177
11		Appendix A	189
	11.1	Taxes on fuels in passenger cars in EU member states	189
	11.2	Physical background of fuel consumption of cars	191
	11.3	Detailed specifications of analyzed cars	195
	11.4	Component costs of propulsion systems	196
	11.5	Economic Assessment	197
	11.6	Net Investment Costs – Compact Class Cars	198
12		Appendix B	200
	12.1	Model calibration	200
		12.1.1 Market shares of technologies	200
		12.1.2 Fleet development	200
		12.1.3 Car characteristics	201
	12.2	Specific Service costs	205
	12.3	Market and fleet penetration	209
		12.3.1 "Business as usual"-Policy & moderate fossil fuel price increase (BAU + Low Price-Scenario)	209
		12.3.2 "Active" Policy & moderate fossil fuel price increase (Policy & Low Price-Scenario)	210
		12.3.3 Fuel tax only & low price – Scenario	211
		12.3.4 Tax on acquisition only & low price – Scenario	211
		12.3.5 "Business as usual"-Policy & substantial fossil fuel price increase (BAU & High Price Scenario)	212
		12.3.6 "Active" Policy & substantial fossil fuel price increase (Policy & High Price-Scenario)	212
	12.4	Shares of vehicle classes in the scenarios	213
	12.5	Fuel supply	214
	12.6	Electricity supply for EVs	216

Abbreviations

μ	expectancy value of the load peak
a	cost of first unit produced
$\bar{a}$	average car age
a_j	diffusion barrier of a technology
AC	alternating current
b_j	technology specific constant
b_l	learning index
b_s	S-curve parameter
BAU	business-as-usual
BEV	battery electric vehicle
C	cost per unit
c_j	technology specific variable
c_w	drag coefficient
CAP	number of cars in the fleet
CC	capital cost
CC_{net}	net capital cost
CC_{sp}	specific capital costs
CD	convention drive
CE	well-to-wheel energy per unit
CF	fixed cost
CI	cost for use of infrastructure
CNG	compressed natural gas
CRF	capital recovery factor
CT	car taxation
D	yearly driving distance of cars
DC	direct current
d_{EV}	distance driven in electric mode per day(BEVs & PHEVs)
DOD	depth of discharge
DT	depreciation time
E	energy demand
E_{in}	energy input
E_M	demand for transport mode
E_{out}	energy output
EC	energy consumption
EC_{CAP}	cumulative energy consumption in the fleet
EC_WTW_CAP	cumulative well-to-wheel energy consumption
EC_{EV}	electricity consumption of EVs and PHEVs in electric driving mode
E_{EV}	cumulative daily electricity consumption
E_M	demand for a transport mode
EM	electric machine

EV	electric vehicle
F	service level of cars
FC	fuel cost
FC	fuel cell
FC-PHEV	fuel cell plug-in hybrid electric vehicle
FCV	fuel cell vehicle
FP	gross fuel price
FP_{net}	net fuel price
FT	fuel tax
GDP	gross domestic product
GE	well-to-wheel greenhouse gas emissions per unit
GHG	greenhouse gases
GHG_CAP	cumulative greenhouse gas emissions
GHG_WTW_CAP	cumulative well-to-wheel greenhouse gas emissions of the fleet
h	index for fuels
HEV	hybrid electric vehicle
i	index for vehicle class
IC_{car}	net investment costs of the car
IC_{CD}	capital cost of conventional drive cars
IC_{comp}	net investment costs of a component
ICE	internal combustion engine
INS	insurance cost
J	technology
j	index for technology
k	number of technology option
LCA	life-cycle analysis
Li-Ion	Lithium Ion
LPG	liquified petroleum gas
LR	learning rate
m	vehicle mass
n	index for vehicle cohort
NEDC	New European Driving Cycle
NiCd	Nickel Cadmium
NiMH	Nickel Metal Hydrid
NOX	nitrogen oxide
NPV	net present value
OC	non fuel operational cost
p	fuel price
P	vehicle power
$P_{cum\text{-}max\_theor}$	theoretic maximum load of electric vehicles
p_{ju}	share of technology in a user group
p_M	price of transport mode
P_{plug}	charging power
p_u	share of a user group
P_{cum_EV}	cumulative load caused by electric vehicles charging

PHEV	plug-in-hybrid electric vehicle	
p_M	price of the mode of transport	
PM	particular matter	
PMSM	permanently magnetized synchronous machines	
q	progress ratio	
r	interest rate	
R&D	research and development	
RES	renewable energy sources	
RES-E	electricity from renewable energy sources	
REX	range extender	
S	energy service	
s	weibull parameter	
SC	service cost of passenger car transport	
SC_{ref}	reference technology	
se	service level	
SHEV	series hybrid electric vehicle	
SOC	state of charge	
SR	survival rate	
SUV	sports utility vehicles	
SZ	survivors of a vehicle cohort	
T	taste	
TA	tax on acquisition	
TCO	total cost of ownership	
TO	tax on ownership	
TTW	tank-to-wheel	
u	user group index	
v	distribution of sold vehicles around the average (in terms of specifications)	
VAT	value added tax	
WTP	willingness-to-pay	
WTT	well-to-tank	
WTW	well-to-wheel	
x	number of produced units	
xr	discrete random variable	
Y	Income	
Z	number of cars newly registered	
z	market share of a technology	
z_{j_cum}	share of a technology in the entire car market	
Z_{SCRAP}	cars scrapped per year	
z_e	percentage of electricity charged in the evening	
z_{j_D}	diffusion curve	
z_m	percentage of electricity charged in the morning	
α_{FP}	fuel price elasticity (fleet)	
α_{IC}	elasticity with respect to fixed costs (fleet)	

α_Y	income elasticity (fleet)
β	elasticity of income
β_{FP}	fuel price elasticity (service level)
β_{IC}	elasticity with respect to fixed costs (service level)
β_Y	income elasticity (service level)
ΔIC_{econ}	economically justified additional cost of hybridisation
ΔIC_{estim}	estimated additional cost of hybridisation
Δz	growth in market share
η	efficiency
η_{cg}	charging efficiency of electric vehicles
λ	weibull parameter
σ	standard deviation if the load peak
ω_{FP}	fuel price elasticity (driving distance)
ω_Y	income elasticity (driving distance)

1 Introduction

1.1 Motivation

Cars play a major role for passenger mobility on a global level today and their relevance is expected to increase in the coming years and decades (WBCSD 2004) (World Energy Council 2007). The dynamic motorisation process in the twentieth century has led to a high degree of individual mobility and flexibility in developed countries, which has strongly affected the life-style of these societies. This mobility concept has become a global paradigm and developing countries make strong efforts to reach a comparable level of mobility.

However, in the last decades the negative consequences of this development have become more and more evident. Today motor vehicles are associated with various negative effects. Two of the major problems passenger car transport will have to face in the coming years are: emissions of pollutants and greenhouse gases (GHG) and fossil fuel dependence

Emissions: Today, propulsion systems of passenger cars are almost exclusively based on internal combustion engines that are fired with hydrocarbon fuels. The combustion process causes both pollutant and greenhouse gas emission. Even though pollutant emissions have been reduced considerably in the last years through improvements of the combustion process and by applying advances exhaust gas after treatment the problem is still not completely solved (Helmers 2009). Especially in urban areas with high traffic densities emissions remain a serious problem that is calling for either regulative or technological solutions.

Due to their reduction commitments set by the Kyoto protocol greenhouse gas emission became increasingly relevant for passenger car transport. In Austria the transport sector has shown the strongest growth in GHG emissions among all other sectors since 1990 (Schneider & Wappel 2009).

There are mainly three ways to reduce greenhouse gas emissions in transport. The simplest option is to reduce the cumulative kilometres driven in the country. Another option is to improve the efficiency of the transport means. The third approach is to use less carbon intense fuels. This thesis will mainly focus on the last two options in the case of passenger car transport. Since greenhouse gas emissions are directly linked to the efficiency of the cars, improvements can directly lead to their reduction. In fact efficiency of passenger cars in Austria has improved in the past years, but these effects have been offset by the growth of the car fleet and the resulting growth of cumulative driving distance (Meyer & Wessely 2009).

Greenhouses gas reduction is also approached by blending biofuels in order to reduce the greenhouse gas intensity of fuels. However, in Austria blending rates are still too low to show significant impact on total GHG emissions of the transport sector (see (Schneider & Wappel 2009)).

Fossil fuel dependence: today passenger cars strongly rely on crude oil based fuels above all gasoline and diesel. Apart from the above mentioned greenhouse gas problem the high import dependence of these fuels is another severe problem for most developed countries. The fact that a major part of the crude oil is imported from a few politically unstable regions is aggravating this problem and has led to a political dependence. The instability of the supply situation is reflected in the historic fluctuations of the crude oil price. Due to its high crude oil dependence road transport is strongly affected by fluctuations of oil price.
In the future oil demand is expected to increase while conventional reserves are decreasing (IEA 2009b). This combination will certainly drive crude oil prices and thereby affect transport costs. The measures that are taken to escape the problems associated with fossil fuel dependence are similar to the measure to avoid GHG emissions. Firstly it is tried to improve the efficiency in order to reduce energy demand and secondly it is tried to introduce alternative fuels to diversify fuel supply.

Today there are legitimate expectations that alternative vehicle propulsion technologies together with alternative fuels could lead the way out of this problematic situation. The European Union is driving an effort to enforce the use of biofuels in order to reduce emissions and dependence on fossil fuels (European Parliament & European Council 2003). However, the potential for substitution is limited and there are concerns, whether the use of fertile land for the production of transport fuels for motor vehicles is justified.
Another approach is the electrification of the vehicle propulsion system. Hybrid and electric cars are in the spotlight today. Some car manufacturers already offer hybrid vehicles in their portfolio and several others are expected to follow within the next years. Furthermore, there are an increasing number of small car manufacturers trying to enter the market with electric cars. Even some of the leading carmakers are announcing the introduction of pure electric cars in the years to come (see (Brunner et al. 2010) (Foley et al. 2010)). It remains to be seen whether the promise of electric vehicles will be converted into commercial success on a large scale.
Yet, besides their ecological advantages (zero emissions) and their superior efficiency electric cars have serious deficits. Their driving ranges are much

lower than the ranges of conventional cars and refuelling is slow. Furthermore, their costs are still too high to address the mass market today. Hybrid cars don't face the driving range and refuelling problems, but their economic success is still highly dependent on the specific framework conditions.

Hybrid electric vehicle (HEV) is an umbrella term that includes a variety of systems with different degrees of electrification: there are systems that are still closely related to conventional systems like micro and mild hybrids, there are systems where the engine and the electric machines equally contribute to the propulsion of the car like full hybrids and there are systems that are closely related to pure battery electric cars like plug in hybrid electric vehicles (PHEV).

The term electric car usually refers to cars that are using an electric drive system with an electro-chemical battery for electricity storage. In addition to these battery-electric vehicles (BEV) there are also electric propulsion systems that are using other technologies to store the energy on board. For example there are series hybrids that use an internal combustion engine to generate electricity on board of the car by driving a generator while the actual propulsion system is electric. The main idea is to store the energy on board in another form with better storability. The same idea is followed by fuel cell vehicle (FCV) where a fuel cell is used to produce electricity from hydrogen on board.

All these propulsion technologies are associated with the development of vehicle powertrain electrification. However, the technical and economic potential of each technology and consequently their role in future passenger transport remains uncertain.

1.2 Objective

This thesis will provide a closer view on the performance of electrified powertrain systems from a technical and economic perspective.

The global objective of this thesis is to analyse how hybrid and electric propulsion technologies can contribute to the reduction of energy consumption and greenhouse gas emissions of the passenger car fleet in general and in the particular case of Austria.

In pursuit of this global objective, the thesis addresses the following questions:

o   What is the economic performance of electrified propulsion technologies today and what are their perspectives for the future?
o   What are the crucial factors for the spread of hybrid and electric cars?
o   Within what time frames hybrid and electric cars can attain considerable market shares?

o   What role can policy play to encourage the spread of these cars and to improve the efficiency of the sector as a whole?
o   What are the main drivers of energy consumption and greenhouse gas emissions in passenger car transport?
o   How will large scale introduction of hybrid and electric cars affect the primary and final energy demand of the car fleet?
o   What is their potential to reduce GHG emissions within the transport sector?

1.3   Method of approach

The objectives of this thesis are pursued by a two-stage methodological approach. First a techno-economic assessment of electrified powertrain systems is performed. Secondly the effects of large scale market introduction is analysed in a model-based analysis. Thereby, a model of the Austrian passenger car fleet is used to develop scenarios for the time frame 2010-2050:

- Assessment of electrified propulsions technologies from a technical and economic perspective;

The technical and economic status of 2010 is determined and the key factors for their economic competitiveness are identified. In order to provide a basis for the energy modelling the cost development of all technologies is estimated for the time frame 2010-2050 in consideration of technological progress and global fossil fuel prices.

- Development of an **energy economic model of the Austrian passenger car fleet** capturing the major developments and interrelations that affect energy consumption and greenhouse gas emissions;
o   The main drivers of energy consumption and greenhouse gas emissions in passenger car transport are identified and implemented in the model in accordance with the theoretic framework of energy economics.
o   Dynamic aspects like technological change (e.g. technological learning) and technological diffusion are captured.
o   The impact of both, policy instruments and fossil fuel prices is modelled.

- Scenarios with different political and economic framework conditions for the time frame 2010-2050;

The scenario results include the fleet development in terms fleet size, vehicle use, properties and technologies of cars as well as energy consumption, energy carriers and greenhouse gas emissions on well-to-wheel (WTW) basis.

1.4 Main literature

To meet the objectives of the thesis methodical inputs have been derived from the following references:

For the techno economic assessment several international publication have been analysed: The analysis of (Kalhammer et al. 2007) (MIT 2008) includes a techno economic assessment of electric vehicle powertrain systems for North America. (IER 2009) and (Wietschel & Dollinger 2008) performed comparable analyses for European conditions.
For the cost estimations of vehicles and components inputs from various reports have been consulted e.g. (EUCAR et al. 2006), (Passier et al. 2007), (Matheys & Autenboer 2005),

In the course of the development of the scenario model other modelling concepts and approaches were studied e.g. (MIT 2008); (Zachariadis 2005); (Fulton et al. 2009); (Ceuster et al. 2007).

The transport economic definitions given in (Button 2010) have been valuable inputs for the applied approach to model energy demand in the passenger car sector. Also the theoretic considerations of energy services and service levels by (Haas et al. 2008) have strongly influenced the method energy demand is modelled.

The theoretical description of the rebound effect given in (Sorrell 2009) as well the empirical analysis of the effect in the case of passenger cars given by (Schipper et al. 2002) have pointed out the need to consider these effects in the model.

Some key advices for the parametrisation of the top-down transport demand model have been derived from (Dargay & Gately 1999), (Johansson & Shipper 1997) and (Goodwin et al. 2004)

(G. Erdmann & Zweifel 2008), (Jaccard 2009) and especially (Axsen et al. 2009) give a comprehensive overview on the implantation of choice models in bottom-up models respectively in models that combine bottom-up and top down aspects.
(Christidis et al. 2003) demonstrates how passenger car fleets can be modelled on a bottom-up basis.

(Nakicenovic 1986) analyzes technological changes in automobile history and gives an impression of the dynamics of technology diffusion processes in this field. (Grübler 1998) addressed key aspects of the technological diffusion and technological learning of energy technology. (McDonald & Schrattenholzer 2001) give an empirical review on learning parameters for energy technologies.

1.5 Structure of the Thesis

An overview on the status of passenger car transport in Austria is given in chapter 2. The development of energy demand, energy carriers and greenhouse gas emissions of the Austrian passenger car fleet as well as the development of the car fleet in terms of technologies, vehicle characteristics and user pattern are illustrated. Finally, the main policy instruments for passenger cars in the EU are presented and a closer view on the Austrian policy framework is taken.

Chapter 3 explains the energy and transport economic background of energy demand in the passenger car sector. It describes some key principles that have to be considered in transport related energy models and thereby, it provides the necessary theoretic basis for the model described in chapter 6.

Chapter 4 looks into the theory of technological change with special focus on propulsion technologies for passenger cars. By explaining some key terms like technological life cycles and diffusion, technological learning and diffusion barriers it imparts the theoretic background of some key elements of the methodical approach described in chapter 6.
By giving a short retrospect on past technological trends and by analyzing the role of major stakeholders it is tried to identify drivers and barriers for the diffusion of alternative propulsion technologies.

Chapter 5 provides the techno-economic assessment of electrified propulsion technologies. For a better understanding of the functioning of vehicle powertrain electrification some basic principles of fuel consumption of cars are explained. After a brief description of different hybrid technologies and their key components their cost is estimated for the technological and economic status of 2010.
To analyze the impact of fuel prices and costs of key components on total cost of the propulsion systems sensitivity analyses are performed. Finally, the cost development 2010-2050 is estimated in scenarios. Thereby, the future development of battery system costs is estimated through learning curves.

In chapter 6 the scenario model of the Austrian passenger car fleet is presented. This includes a global overview on the applied approach and a detailed description of the methodological implementation. The main aspects described are the modeling of market shares of technologies, the modeling of shifts in demand and service level of passenger car transport and the bottom-up modeling of the fleet.

Chapter 7 introduces the scenarios that are developed with the model. The scenario assumptions include different policy schemes and different fossil fuel price developments. The first part of the scenario results is presented, namely the market and fleet penetration of vehicle propulsion technologies as well as the development of the car fleet in terms of quantities, and average characteristics of cars sold.

Chapter 8 focuses on energy demand and greenhouse gas emissions in the scenarios. First the scenario assumptions concerning energy supply are presented which include different rates of biofuel blending and different sources of electricity. Based on these scenario settings the second part of the results is presented including energy consumption, energy carriers and greenhouse gas emissions. Furthermore, a brief estimation of the effects of cumulative EV charging on the Austrian electricity load curve is made, based on the fleet penetrations derived from the scenarios.

In chapter 9 conclusions and recommendations are derived.

2 Energy economic parameters of passenger car transport in Austria

In Austria today the car is the most important mean of passenger transport and its relevance is still increasing. Figure 2-1 indicates the importance of the passenger car to cover individual transport demands: Today the car is the preferred mean of transport for almost 70% of total transport trips in Austria.

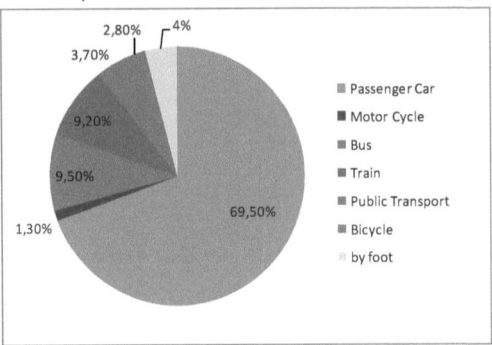

Figure 2-1: Share of Transport Media in Austrian Passenger Transport (Source: Federal Ministry of Environment 2009 (Schneider & Wappel 2009))

This chapter gives an overview on the energy demand of passenger car transport and its main drivers. Thereby the most relevant indicators for transport energy demand and their historic development are presented. This includes data on transport intensities, fleet statistics, technological trends, efficiency, energy consumption and emissions of the passenger car fleet in Austria. This data also represents a major input for the energy economic model that will be described later in this thesis (chapter 6 to 8).

The chapter is concluded with a brief overview on political framework conditions of passenger car transport in the European Union and in Austria.

2.1 Energy Demand of passenger cars in Austria

With more than 60 % of the global crude oil supply the transport sector is currently the biggest consumer of crude oil products (IEA 2009a). Also in Austria the energy supply of the transport sector relies almost completely on oil based energy carriers. With more than 90 % of all transport energy, road transport is by far the biggest consumer within the sector (Herry et al. 2007).

In Austria the main final energy carriers for road transport are diesel and gasoline fuels. Since 1990 the demand for diesel fuel has strongly increased driven by two main developments: firstly by the trend toward diesel passenger cars between 1990 and 2005 and secondly by the price difference of diesel between Austrian and its neighbour states that has attracted many international trucks to refuel in Austria and has caused considerable additional consumption (see chapter 2.2).

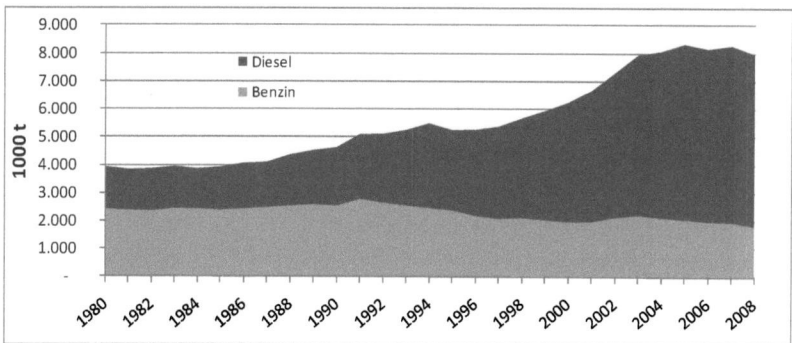

Figure 2-2: Energy Carriers for road transport in Austria (Data Source: (Fachverband Mineralölindustrie 2010a))

Both diesel and gasoline are blended with biofuels in Austria. Following the EU directive 2003/30/EG they are blended with 5.75 % of biofuels, which means biodiesel in the case of diesel and bioethanol for gasoline (Winter 2008). Other alternative fuels, like compressed natural gas (CNG) or liquefied petroleum gas (LPG), are less relevant for road transport in Austrian and are negligible in the gross energy supply of the sector (Fachverband Mineralölindustrie 2010a).

2.2 Emissions of passenger cars in Austria

The transport sector is one of the biggest emitters of greenhouse gases and air pollutants. It causes 32 % of greenhouse gas emissions in Austria. Furthermore, it is the biggest emitter of nitrogen oxide (NOX) (64 %) and causes a considerable share of particular matter (PM) emissions (20 % of PM10, 25 % of PM2,5 and 18 % of PAK-Emissions) which are especially critical since they are often emitted in urban areas (Pazdernik et al. 2009).

In 2007 road transport accounted for 26.7 % of total emissions in Austria (passenger cars: 15.2 %; commercial vehicles: 11.5 %. Between 1990 and 2007 road transport related greenhouse gas emissions (GHG) have increased by 73 % in Austria. Emissions of passenger cars increased by 44 %, while emissions of commercial vehicles increased by 138 % (Schneider & Wappel 2009). GHG emissions are directly linked to the consumption of fossil fuels and therefore show a similar development. The increase in emissions is partly caused by fuel exports through foreign motor vehicles refuelling in Austria. However, a considerable share of emission growth has also been generated by the growing demand of the domestic fleet of passenger cars and commercial vehicles (see Figure 2-3).

The evolution of GHG emissions illustrated in Figure 2-3 also expresses the shift from gasoline to diesel cars in the passenger car fleet. This trend did in fact improve the efficiency of passenger cars to some extent but it had negative effects on the emissions of some air pollutants. Diesel cars have significantly higher emissions of nitrogen oxide and particular matter than gasoline cars. In Europe the legislation for air pollutants distinguishes between gasoline and diesel cars allowing the latter higher emission thresholds for particular matter and nitrogen oxides (Wallentowitz et al. 2010).

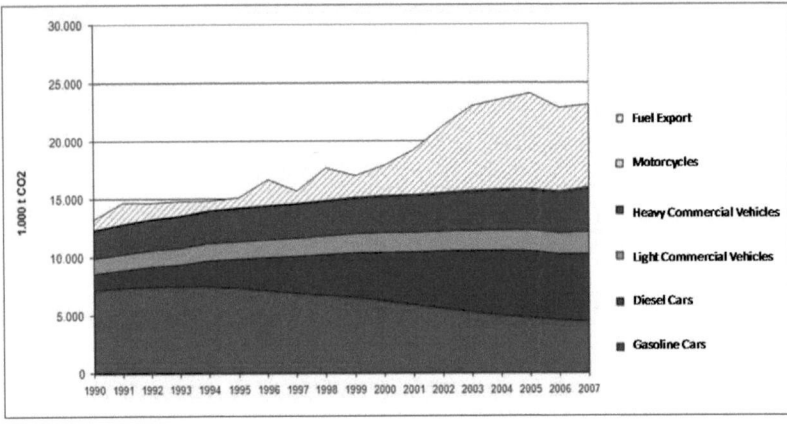

Figure 2-3: Greenhouse gas emissions caused by road Transport in Austria (Source: Federal Ministry of Environment 2009)

2.3 The Austrian passenger car fleet

The growing importance of individual motorised transport in Austria is reflected in the growth of the national passenger car fleet. From 1970 to 2009 the fleet has almost quadrupled, reaching 4.3 million vehicles. For a long time gasoline engines have been the dominating propulsion technology in the fleet. In the last two decades there has been a major shift from gasoline to diesel engines. This trend was mainly driven by the lower tax on diesel and the development boost in diesel engine technology that made the technology more attractive to consumers. Also policy makers considered diesel cars an appropriate measure to reduce fuel consumption and cut greenhouse gas emissions (GHG). They even adapted exhaust gas legislation with higher emission thresholds for diesel engines to facilitate their diffusion (Helmers 2009).

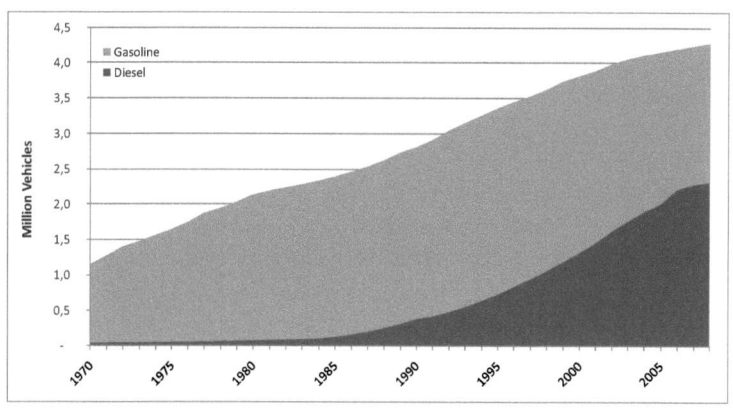

Figure 2-4: Development of the Austrian Passenger Vehicle Fleet 1970-2008 (Data Source: (Statistics Austria 2009b))

2.4 Trends in passenger car sales

The automotive industry is one of the most important industries in the world. Globally more than 60 million motor vehicle are produced per year among those more than 50 million are passenger cars and light trucks (VDA 2010).

Between 1980 and 2008 the Austrian car market has grown from around 200 000 cars year^{-1} to around 300 000 cars year^{-1} but there have always been fluctuations from one year to the other (see Figure 2-5). The fact that growth in sales was slower than the growth of the car fleet indicates that cars tend to remain longer in the fleet. The average age of the Austrian fleet was 7.4 years in 2006, which is among the lowest in the European Union (ACEA 2010).

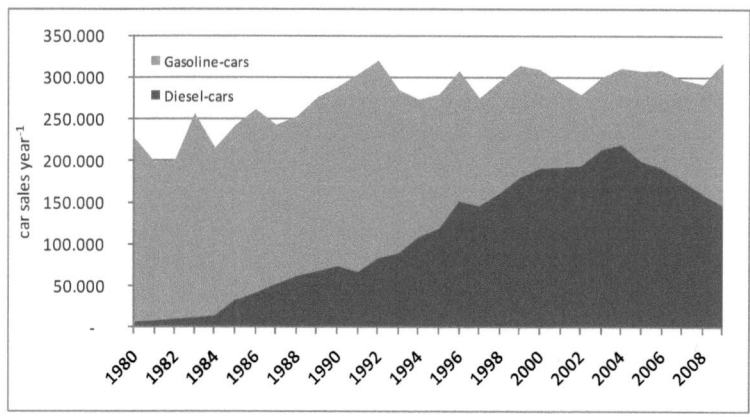

Figure 2-5: Development of Passenger Vehicle Sales in Austria 1980-2009 (Data Source: Statistics Austria)

2.4.1 Propulsion technologies

After the strong trend toward diesel cars starting in the mid 1980ies there has been a turnaround in this trend in the last years (see Figure 2-5). The main driver for the comeback of the gasoline engine was the efficiency improvement that could be achieved at this technology. Measures like downsizing and turbo charging significantly improved the efficiency of gasoline engines. (see chapter 2.4.2).

The Austrian car sales of 2009 illustrated in Figure 2-6 shows that cars with alternative propulsion technologies still play an inferior role in the passenger car market accounting for less than 1% of 2009s sales. In fact hybrids have experienced a continuing growth in the last years but their absolute number of sales of 1055 (2009) is still very low. The same applies for CNG cars of which only 500 were sold in 2009.(Statistics Austria 2009b)

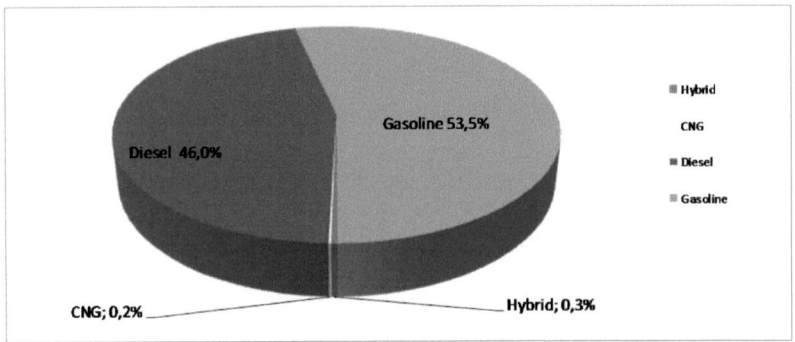

Figure 2-6: Shares of Technologies within Austrian Passenger Vehicle Sales 2009 [Source. Statistics Austria]

2.4.2 Curb weight and engine power

As indicated in Figure 2-7 the average curb weight of cars sold in Austria has strongly increasing in the last decade. From 2000 to 2008 it has increased by about 150 kg with diesel cars showing a disproportionately strong increase (see Figure 2-8).

The main driver of the growing average mass has been the increasing number of comfort and security features in all cars and the rising market shares of SUV cars. Keeping the same driving dynamics for these heavier vehicles requires stronger engines which has also lead to a considerable increase of average engine power especially for diesel cars. What is noticeable is the fact that the average power of gasoline cars did barely increase in this time frame. This can be explained by the fact that gasoline engines remained the dominating technology for small cars while diesel engines became the first choice for bigger cars such as the popular sports utility vehicles (SUV) (Pötscher 2009).

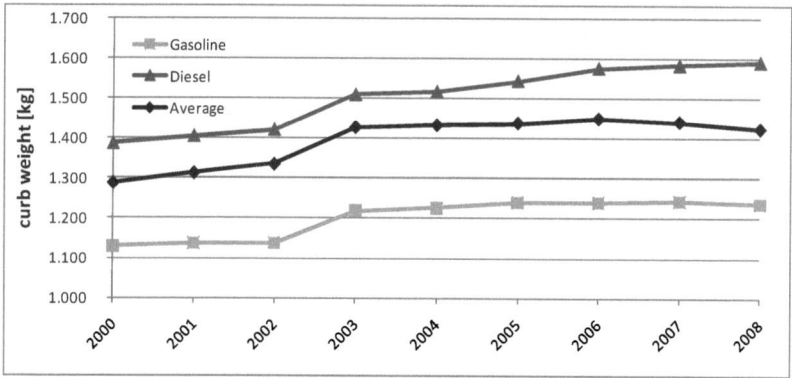

Figure 2-7: Average curb weight of passenger cars sold in Austria (Data Source: Federal Ministry of Environment (Pötscher 2009))

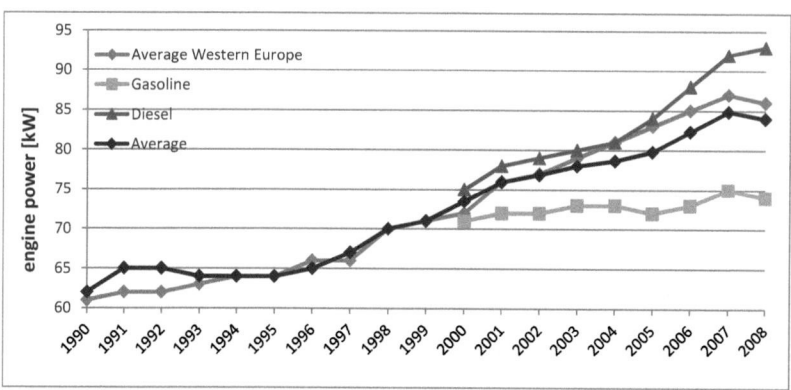

Figure 2-8: Engine power of passenger cars sold in Austria (Data Source: Federal Ministry of Environment (Pötscher 2009))

One example for the tendency toward higher mass and engine power is the historic development of the popular Volkswagen model Golf (see Table 2-1). The data of the historic VW Models show that even though the car has always been in the same vehicle segment (Compact Class/Golf-Class), the specifications changed dramatically. For gasoline models the vehicle mass grew by almost 500 kg from the first generation to the sixth. The engine power almost doubled in the same time. The same development was found by (MIT 2008) who analysed the curb weights of Toyota corolla models sold in the USA between 1990 and 2006. The development of curb weight of some of the most important car models sold in Europe show that this long term trend has even accelerate in the last years (see (Berger et al. 2009)). This shows that the efforts taken by car manufacturers to reduce vehicle mass by use of light-weight materials have been offset by the weight increase caused by additional comfort and safety features.

Table 2-1: Specifications of historic VW Models   Source: (Helmers 2009), slightly adapted)

	Construction Year	Type	Motor		Weight	Top Speed	Fuel Consumption
			Displacement l	Power hp	kg	km/h	l/100km
Gasoline	1948	Kaefer	1.1	24.5	600	100	7.5
	1973	Kaefer	1.2	42	760	115	7.5
	1978	Golf I	1.1	50	750	140	8.3
	2008	Golf V	1.6	102	1173	184	7.4
	2009	Golf VI TSI	1.4	122	1241	200	6
Diesel	1978	Golf I	1.5	50	805	140	6.5
	1993	Glof III (Ecomatic)	1.9	64	1115	155	5.5
	2008	Golf V (Blue Motion)	1.9	105	1200	190	5
	2009	Golf VI	2	140	1322	207	5.4

2.4.3 Fuel consumption and emissions

Figure 2-9 and Figure 2-10 show the average fuel consumption and the corresponding greenhouse gas emissions of cars sold in Austria from 1990 to 2008. Between 1990 and 2000 emissions of both gasoline and diesel cars have decreased slightly. The trend towards diesel cars in this time period lead to a reduction of average fuel consumption and GHG emissions. From 2000 to 2008 there were different tendencies for gasoline and diesel cars. While average fuel consumption of gasoline cars decreased continuously average consumption of diesel cars remained more or less constant and even increased between 2005 and 2007. The main reason for this development is the fact that average mass and power of diesel cars increased significantly and the diesel engine became the dominating propulsion system for large and heavy cars like SUVs. Considering the entire time range between 1990 and 2008 fuel consumptions and emission were reduced despite of the fact that average mass and power have increased considerably in the same time. This indicated the efficiency improvement of internal combustion engines that have been achieved in this time frame.

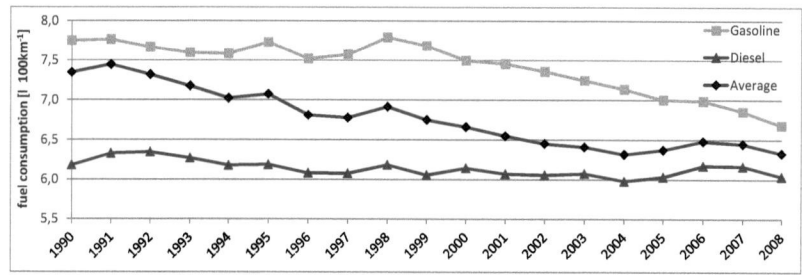

Figure 2-9: Average fuel consumption of cars sold in Austria (Data Source: (Odyssee 2010))

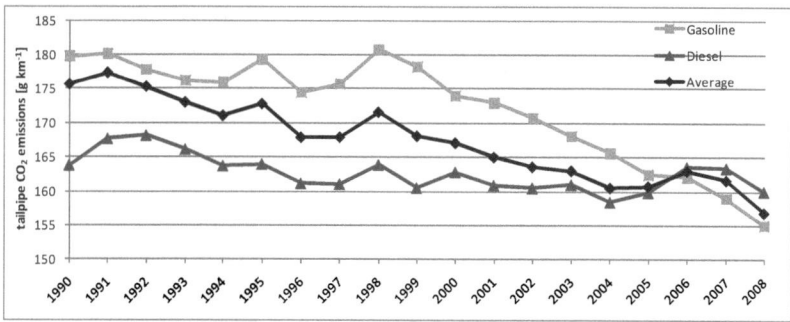

Figure 2-10: Average Greenhouse Gas Emissions of Austrian car sales (Data Source: (Odyssee 2010))

2.5 User pattern of passenger cars in Austria

For an energy-economic analysis of passenger car transport, the average intensity vehicles are used is highly relevant. An important parameter in this context is the average yearly driving distance of cars in fleet. In Austria the average yearly kilometrage of cars was 13 500 km in 2008 (see Figure 2-12) (Statistics Austria 2009a). There was a significant difference in user intensity between diesel and gasoline cars. The average kilometrage of diesel cars was 15 200 km while the one for gasoline cars was only 11 300 km. This can be explained by the simple fact that diesel cars are usually used by commuters due to their lower fuel costs. There is also a clear difference in user intensities of first and second cars in a household where the second cars are used much less intense than the first cars (see Figure 2-13). The yearly kilometrage is a crucial parameter for the consumer's choice of a vehicle propulsion system. This coherence is also reflected in the market shares of technologies for the first and the second car of a household (see Figure 2-14).

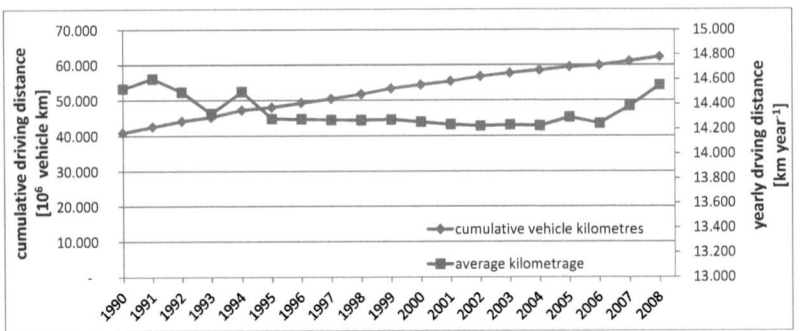

Figure 2-11: Average and cumulative yearly kilometrage of Austrian passenge cars (data source: (Statistics Austria 2010b))

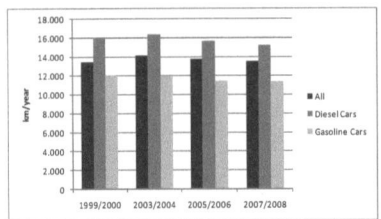

Figure 2-12: Average kilometrage of Austrian passenger cars 2008 (Data source: (Statistics Austria 2009a))

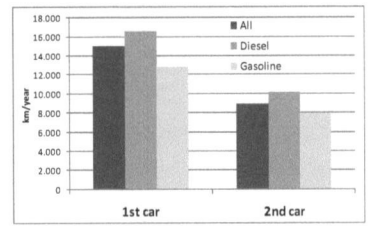

Figure 2-13: Average kilometrage of first and second car of a household 2008 (Data source: (Statistics Austria 2009a))

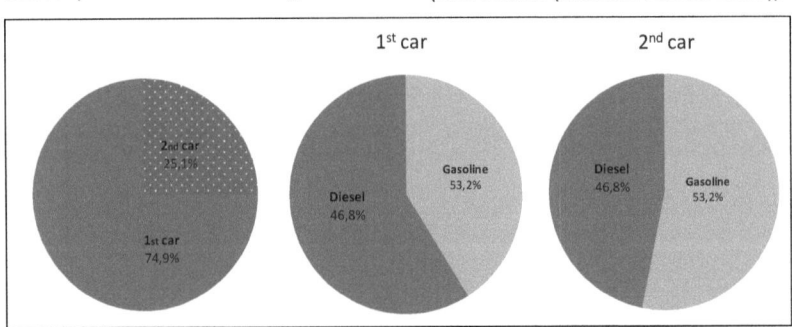

Figure 2-14: 1st and 2nd cars of households (Source: (Statistics Austria 2009a))

Apart from the absolute kilometres driven the detailed pattern of vehicle use is of major interest especially for vehicles using technologies that have to struggle with limitations in driving range like battery electric cars. An electric car not only has limited driving range but also has the problem that it cannot recharge at every petrol station within a few minutes. In Austria more than 90 % of trips done by passenger cars are not longer than 50 km, which points out that the long ranges of up to 1000 km, offered by conventional cars, are not necessary to cover the daily requirements of a typical passenger car user. Figure 2-15 shows the frequency of trips and the kilometres that are driven in these trips. It shows that 90 % of the trips are not longer than 50 km and two third of the kilometrage fall upon trips within this length. The share in kilometrage of trips longer than 200 km is only 8 %, the share of trips longer than 500 km only 1 % (VCÖ-Forschungsinstitut 2009). This underlines that the driving ranges of cars that we are used today, are far above the actual every day needs.

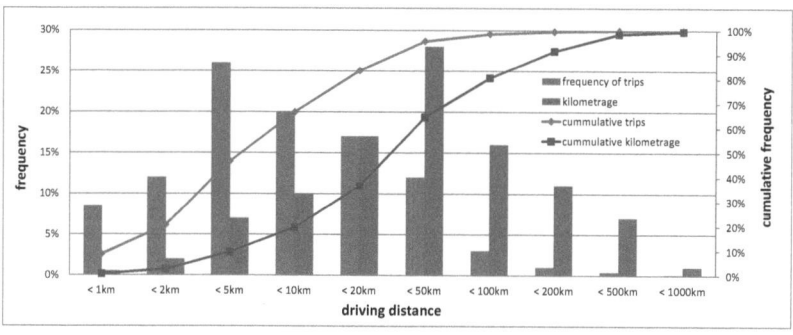

Figure 2-15: Frequency and length of Trips of Passenger Cars in Austria (VCÖ-Forschungsinstitut 2009)

2.6 Policy in the passenger car sector

The political framework for passenger cars mainly consists of two groups of instruments: direct measures like regulatory measures and indirect ones called market based measures. Regulatory measures are applied for example in the case of exhaust gas emission standards and safety standards for cars. Market based instruments include all types of taxes and charges on vehicles, fuels and infrastructure use.

The basic idea of most taxes on passenger cars is to internalise costs caused by external effects (also called external costs) of the transport mode (see chapter 3.8). The most important types of external costs in the case of passenger car transport are infrastructure costs, accident costs and environmental costs (e.g. air pollution, noise, GHG emissions...). In many countries, including Austria, costs of road transport are only partly internalised (Frey et al. 2007).

There are efforts in the European Union toward a better internalisation of external transport costs for all modes through political framework conditions. The main motivations are to make the real costs of transport represented in the price of the service, to improve efficiency and to reduce its environmental impact (Essen et al. 2008).

By internalising external costs policy also affect the economic attractiveness of passenger cars as a transport mode in comparison to other modes. Furthermore, it can influence the average characteristics of the cars in the fleet in terms of weight, power, propulsion technologies and fuels. It is evident that the external costs of transport of a car strongly depend on the technology it is using (e.g. zero emission cars cause lower external costs than conventional cars). Therefore, internalisation of external costs can help to promote efficient cars and clean propulsion technologies. This is why policy can play an important role for the diffusion of alternative vehicle propulsion technologies in the passenger car sector.

Another motive of policy measures in road transport is the reduction of GHG emissions. Austria's greenhouse gas reduction commitment defined by the Kyoto protocol obliged the country to considerably reduce emissions compared to the reference year 1990. While emissions could be reduced in some sectors (e.g. agriculture, space heating) there are others where emissions have increased significantly in the same time frame (e.g. the industry and transport sector) (Schneider & Wappel 2009). This indicates that the measures implemented in the time frame 1990-2010 have been insufficient. A future reduction of GHG emissions will ask for enforced measures in the transport sector with special focus on road transport which is the major emitter.

This chapter will present taxations schemes applied in Europe and will analyse their effectiveness in promoting efficient vehicle technologies. In Europe there are three main types of taxes that affect passenger cars:

- Fuel Tax
- Tax on Acquisition
- Tax in Ownership

2.6.1 Tax on Acquisition/Tax on Registration:

Tax on acquisition or registration tax is paid once when the vehicle is registered for the first time in the country. In Europe there are different schemes to assess this tax. Usually the tax is levied depending on parameters like cylinder capacity, engine power, vehicle mass, fuel consumption or combinations of these. An overview on taxation schemes of vehicles throughout the European Union is given in Table A-1 in Appendix A.

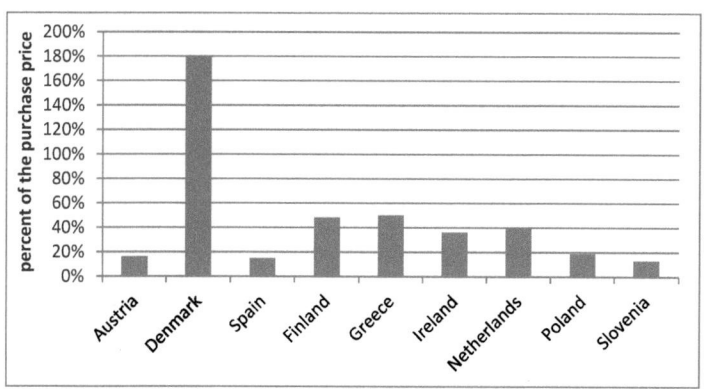

Figure 2-16: Taxes on Acquisition in selected EU Member states (Data Source: (ACEA 2009))

Basically tax on acquisition can be a suitable instrument to internalise external cost of passenger car transport. By setting higher taxes on vehicles that cause higher external costs, the tax can promote efficient and environmentally friendly technologies. However, not all existing schemes are able to tax vehicles correctly by the external cost they cause. For example the cylinder capacity or the price of a vehicle gives no information on its external costs. An adequate approach for vehicle taxation is to use fuel consumption, respectively greenhouse gas emissions, and exhaust gas emissions of the car to quantify the vehicle tax. In the case of taxes on acquisition this approach is applied only by a minority of EU member states, while most countries either have out-dated taxation schemes or no tax at all. The rates of tax on acquisition are very different throughout the European Union. Figure 2-16 shows the maximum tax rates of EU countries, where tax on acquisition is determined as percentage of the car's purchase price.

2.6.1.1 Tax on Acquisition in Austria

In Austria this tax has to be paid just once when the car is registered for the first time in the country and is levied as a percentage of the purchase price. The percentage depends on the car's fuel consumption and is caped with 16 % of the purchase price. Also, there is an additional bonus/malus system on greenhouse gas emissions. When the vehicle's emissions are below or above a certain threshold, the tax is reduced or increased by 25 € g^{-1} CO_2. The upper threshold for the bonus in 2010 was 160 g km^{-1}, the lower threshold for the malus was 120 g km^{-1} (see Table 7-1 in chapter 7). Moreover, there are special deductions for vehicles that use an alternative propulsion system (-500 €) while zero emission vehicles pay no tax on acquisition at all. The Austrian scheme promotes efficient cars and alternative propulsion technologies to some extent.

2.6.2 Tax on Ownership:

The tax on ownership is levied yearly as a fixed tax regardless how many kilometres the car is driven. Within the EU-27 there exist many different schemes for tax on ownership. Most are based on engine power, CO_2 emissions, fuel consumptions or cylinder capacity, but there are also schemes that are based on weight, exhaust emissions and age of the cars (see (Ajanovic et al. 2009)).

To internalise the external cost of transport tax on ownership has to be based on emissions of exhaust gases, greenhouse gases and noises etc. which would also be most favourable when it comes to promoting efficient and environmentally friendly vehicle technologies.

2.6.2.1 Tax on Ownership in Austria

In Austria the rate of this tax depends on the engine power of the vehicle and is paid on a yearly basis. Since vehicles with higher power are usually less efficient the tax has some kind of regulative effect on the efficiency of vehicles sold, but when it comes to comparing systems with the same power there is no differentiating between high and low efficient technologies. Summing up it can be said that the current tax on ownership in Austria has very limited effectiveness in promoting fuel economy, as it gives no direct incentive to choose improved technologies.

2.6.3 Fuel Tax:

Motor vehicle fuels are taxed with different rates in EU countries (see Figure 2-17). Basically the fuel tax is an effective regulatory instrument to influence consumers in terms of their choice of vehicle as well as its use. Since fuel cost is a function of vehicle efficiency, fuel price and the distance driven (see chapter 3.8) an increasing fuel price as a result of higher fuel taxes would induce consumers to either switch to cars with higher efficiency or to reduce their yearly driving distance.

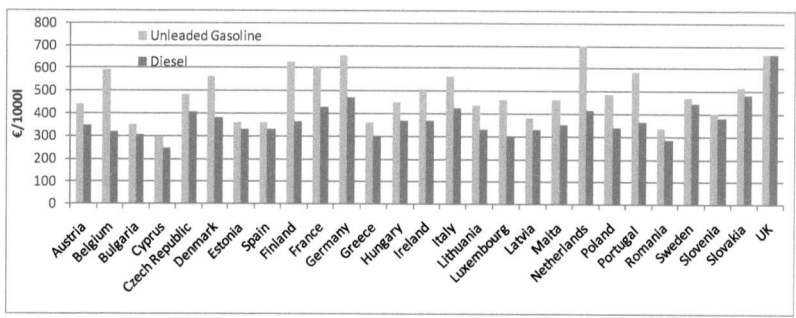

Figure 2-17: Fuel Taxes in EU27 (Data Source: (European Commission 2010))

2.6.3.1 Fuel tax in Austria

The fuel tax in Austria is 0,447 € liter^{-1} on gasoline and 0,347 € liter^{-1} on diesel. Biofuels and CNG are excluded from the fuel tax so far (status 2010) (see Table 7-2 in chapter 7.2).

In comparison to other European countries Austria has average fuel tax rates. However, in comparison to its large neighbour states Italy and Germany the Austrian fuel tax is relatively low. Especially the low tax on diesel fuel makes diesel cars economically attractive (see chapter 5.8). The resulting price difference causes many foreign cars and trucks to refuel their vehicles in Austria when passing through the country. The fuel consumed by these cars causes almost 30 % of the Austrian domestic fuel consumption (Schneider & Wappel 2009). This has positive effects on the Austrian national budget but negative effects on the Austrian greenhouse gas emission balance. The latter could force Austria to adapt its fuel taxation to the level of the European neighbour states.

2.6.4 Emission standards

Apart from taxes there are regulations on emissions of passenger cars. In the European Union there is a common pollution regulation that defines thresholds for the major pollutants of passenger cars. Emission regulation can have a strong impact on technologies and fuels used in the fleet.

For example the EU regulation has different threshold for diesel and gasoline cars, allowing diesel higher emissions of particles and nitrogen oxide. In the USA and Japan where no exception for diesel cars exist, there are almost no diesel passenger cars in the fleets (Helmers 2009). In the future national or regional emission standards can play a key role for the diffusion of low or zero emission technologies.

3 Some theoretical background of energy economics and energy modelling in the transport sector

This chapter provides some theoretic background of the modeling that will be presented later in this work (see chapter 6). It will briefly explain the main concepts and principles the modelling approach is based on and should also provide a better understanding of the key factors and developments that affect energy demand in passenger car transport.

3.1 Transport as an Energy Service

One basic principle of energy economics is the fact that there is no demand for energy but only for energy services (Haas & Wirl 1992) (Haas et al. 2008). The transport service thereby is a function of energy and efficiency of the technology that provides the service:

$$S = f(E, \eta(j)) \tag{3-1}$$

S ... energy service
E ... energy demand
η ... efficiency of a technology j

In terms of passenger car transport this means that consumers don't have demand for fuels but for transport services, which means that they want to travel from A to B. Therefore, the amount of energy required is determined by the demand for transport services and their efficiency. Consequently energy consumption can also be seen as function of energy services consumed and their corresponding efficiency:

$$E = f(S, \eta(j)) \tag{3-2}$$

The resulting equation mainly indicates the principle idea that has been applied to approach the core questions in this thesis: In order to analyse effects of changing framework conditions on energy demand, their effects on transport demand and the efficiency of transport services have been analysed. In the case of the passenger car transport this means that the demand for this mean of transport (expressed in vehicle kilometres or passenger kilometres) and the efficiency of the cars are determined in order to derive total fuel demand.

The following sections will give a brief overview on energy economic aspects of passenger car transport including a short introduction of the theory of transport service demand and transport service levels. Furthermore, some key parameters for transport demand and service level will be explained and it will be illustrated how these parameters can be integrated in energy models for the transport sector.

Thereby, the chapter should provide the required theoretical background for a better understanding of the method of approach that is applied in the model of Austrian passenger car transport which will be presented later in this thesis (see chapter 6).

3.2 Demand for Transport

The demand for transport can be explained by the simple economic demand theory. Demand for a transport mode is basically determined by factors like the price of the transport-mode, income, prices of alternative modes, tastes etc. The demand curve for transport generally can be expressed as follows (Button 2010):

$$E_M = f(p_M, Y, p_1, p_2, ... p_n, T) \qquad (3-3)$$

E_M ... demand for a transport mode
p_M ... price of the mode of transport
Y ... income
p_n ... price of alternative modes of transport
T ... taste

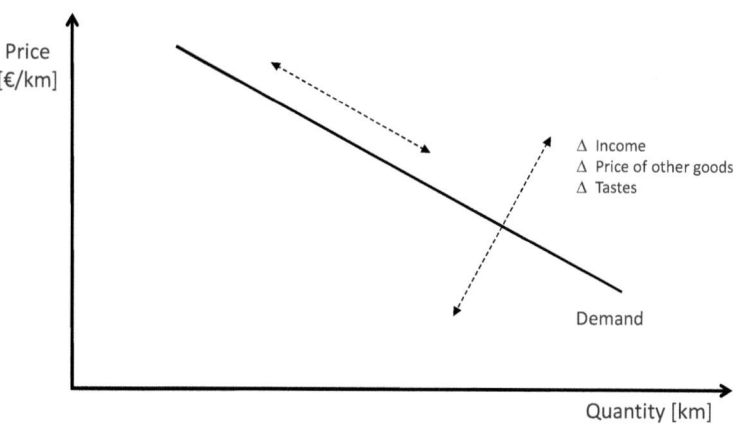

Figure 3-1: Demand Curve for Transport (Source: K Button 2009)

To estimate the effects of the parameters in equation (3-3) on the demand of transport a log linear specification can be used (Button 2010):

$$\ln E_M = \sigma + \beta \ln Y + \gamma_M \ln p_M + \gamma_N \ln p_N \qquad (3\text{-}4)$$

β ... elasticity of income
γ_M ... elasticity of price for the transport-mode
γ_N ... elasticity of price for alternative transport-modes
σ ... constant determining the level

This general theory can now be specified for the particular case of passenger car transport. In a first step it is necessary to clearly define how demand is reflected in measurable parameters in this case.

Changes in the demand for passenger car transport can be expressed in several ways. They influence the use intensity of the cars, reflected in their yearly average driving distance. Consumers react to decreasing income or increasing prices of the transport mode, by planning trips more efficiently or by switching to other transport modes (see chapter 3.5).

Furthermore, demand shifts affect the size of the passenger car fleet. Growing demand for passenger transport leads to growth in the vehicle fleet. The sensitivity of the car fleet to demand shifts depends on the framework conditions in the specific country or region and will be discussed in more detail in chapter 3.4.

3.3 Energy demand

In top down analyses the energy demand is often used as an indicator for the development of overall transport activity. One main advantage of this parameter is its good measurability as it is directly represented in the demand of gasoline and diesel, with usually adequate statistic coverage. However, energy demand can only be used as an indicator for transport activity when efficiency is known (see equation (3-1)).

Energy demand can be used for rough top down estimations within the entire road transport sector. For a detailed analysis of transport demand in the passenger sector the indicator is much too aggregate. In bottom up model historic energy demand statistics can be used to calibrate model settings as it will be showed later in this thesis (see chapter 6.5).

3.4 Car Ownership

The parameter that expresses the car ownership is the so called motorisation, which denotes the number of cars per 1000 inhabitants. Basically, the main parameters that influence the motorisation are the income level and the service cost of passenger car transport. Furthermore, according to the general transport demand theory described above, they are also influenced by the costs of competing transport modes. Another relevant parameter is the infrastructure. Availability and quality of the infrastructure, plays an important role for the attractiveness of a transport mode. Finally, there is a parameter named "tastes" above that also plays a role for car ownership. Apart from pure economic considerations there is also an emotional background of car ownership that will be discussed in more detail later in this work (see chapter 4.6.1).

International analyses have shown that income is the most important parameter for car ownership (Dargay & Gately 1999). It is evident that motorisation in a large scale requires a certain income level in that specific country or region to enable a broader range of consumers to afford a car. Therefore, a minimum income level can be seen as necessary condition for the generation of demand for a specific mode of transport. A comparison of motorisation and demand in different countries points out the strong relationship between these parameters (see Figure 3-2). In developing countries with low per capita GDP car ownership lies below 0.1 cars per capita while in developed countries there are more than 0.5 vehicles per capita. International analyses show that this correlation can be represented by an S-shaped curve. At very low income levels (<5000 $ cap^{-1}) motorization is close to zero and increases only slowly when income increases. At higher levels (5000-15000 $ cap^{-1}) car ownership increases faster and enters a saturation phase starting at high income levels (> 20 000 $ cap^{-1}) (see (Dargay & Gately 1999)).

In literature the demand for passenger cars, respectively the development of the vehicle stock can be modelled as a function of income Y, fuel price p, fixed costs (car taxation) CF and population density G (Storchmann 2005) (Johansson & Shipper 1997):

$$CAP_t = f(CAP_{t-1}, Y, p, CF, G) \tag{3-5}$$

The impact of the different parameters on the development of the vehicle stock can be modelled through elasticities as described by the following equation in a general form:

$$CAP_t = CAP_{t-1} \cdot \left(\frac{Y_t}{Y_{t-1}}\right)^{\gamma_Y} \cdot \left(\frac{P_t}{P_{t-1}}\right)^{\gamma_P} \cdot \left(\frac{CF_t}{CF_{t-1}}\right)^{\gamma_T} \cdot \left(\frac{G_t}{G_{t-1}}\right)^{\gamma_G} \tag{3-6}$$

CAP ... vehicle Stock
Y ... Income

P ... running cost
CF ... fixed cost
G ... population density
y_Y ... income elasticity
y_P ... elasticity with respect to running cost
y_T ... elasticity with respect to fixed cost
y_G ... elasticity with respect to population density

As mentioned above the income usually has the strongest impact on motorization and the vehicle stock. The income elasticity indicates how changes in income (usually represented through the GDP capita^{-1}) affect the development of the car stock. As indicated mentioned above, the parameter varies strongly depending on the absolute income level of the country or region (Lescaroux & Rech 2008). This coherence can be described by the Gompertz function which turned out to be appropriate model for the relationship of income and motorization (Dargay & Gately 1999) (J. Dargay et al. 2007). This model explains why countries with high income have considerably lower income elasticities of motorization than countries with lower income.

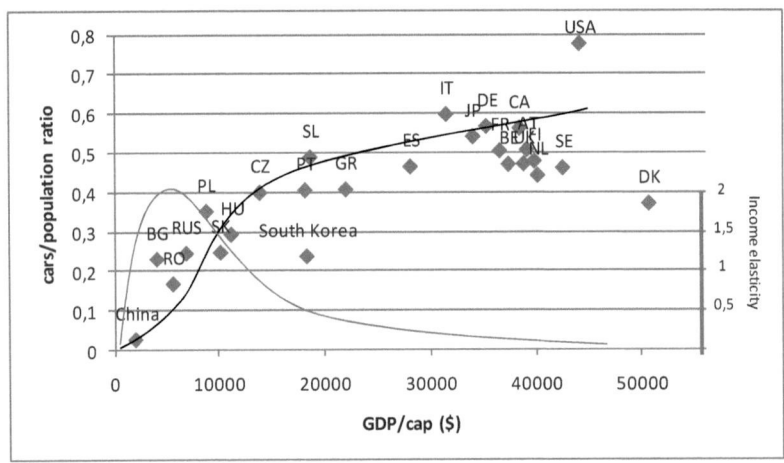

Figure 3-2: Car Ownership 2006 in selected countries (data source (ACEA 2010) & schematic illustration of car ownership and income elasticity (according to (Dargay & Gately 1999))

The illustration of income and car ownership in Figure 3-2 also indicates that there are other factors than income that affect car ownership. Countries like South Korea, Denmark and the USA differ from the global trend. For those countries the other parameters like fuel price, car taxation and population density play an important role. Generally there seems to be a common saturation level of 0.5 cars cap^{-1}. The considerably higher vehicle ownership of the USA might be explained by the lower fuel prices and by the low population density. In the case of South Korea, which has a relatively low vehicle ownership compared to its income level one reason might be the high population density together with the geographically isolation. Denmark is also a special case: The country has a very high income level but a relatively low motorization. One explanation for this contradiction might be the extremely high taxation on cars in this country (see chapter 2.6).

3.5 Car use

Another parameter that is reflecting the demand for passenger car transport is the intensity the cars are used, expressed in kilometers per year. Similar to car ownership the main parameters that are considered to model the mean driving distance per year are income, fuel price, taxation and population density (Johansson & Shipper 1997). Consequently, the yearly driving distance of cars can be defined by the following general equations:

$$D_t = f(D_{t-1}, Y_t, p_t, CT_t, G_t) \qquad (3\text{-}7)$$

$$D_t = D_{t-1} \cdot \left(\frac{Y_t}{Y_{t-1}}\right)^{\beta_Y} \cdot \left(\frac{p_t}{p_{t-1}}\right)^{\beta_P} \cdot \left(\frac{CT_t}{CT_{t-1}}\right)^{\beta_T} \cdot \left(\frac{G_t}{G_{t-1}}\right)^{\beta_G} \qquad (3\text{-}8)$$

D ... mean driving distance of cars
Y ... Income
p ... fuel price
CT ... car taxation
G ... population density
β_Y ... income elasticity
β_P ... fuel price elasticity
β_T ... elasticity of car taxation
β_G ... elasticity of population density

3.6 Service Level in passenger car transport

As explained in beginning of this chapter energy demand is determined by the demand for an energy service and the efficiency of the transport mode that is used to provide the service. The efficiency of the transport mode is determined by the technology that is used but also by the transport service level. The transport service level defines the quality the transport service is provided.

Consequently, the service level is also affecting the energy demand of the transport mode. Referring to equation (3-2) the effect of transport service level on energy demand can be described as follows:

$$E = f(S, \eta(j, F)) \quad\quad (3\text{-}9)$$

E ... energy demand
S ... service demand
η ... efficiency of passenger car transport
j ... technology
F ... service level

In passenger car transport a higher service level means that the same distance is travelled with a more powerful or more comfortable car. Therefore, the average transport service level in a country is reflected in the average mass and engine power of cars. For example at high service cost levels, as a consequence of high fuel prices or high taxes, consumers tend to choose smaller cars with lower engine power – an effect that is also been reflected in Austrian sales statistics (see chapter 2.4).

Thus changes in framework conditions that are affecting the specific service costs also affect the consumer's behaviour when choosing a vehicle category. Just like car use and car ownership the service level is affected by parameters like income, fuel prices, taxation, etc. Estimations in past analysis showed that the fuel price has the strongest impact on fuel consumption (Johansson & Shipper 1997) which, according to the aforementioned theory, can also be seen as an indicator for effects on the service level.

By affecting the efficiency of the cars, the service level has strong impact on the overall energy consumption of the fleet and has to be considered for a correct capturing of the impact of changing framework conditions on energy consumption and GHG emissions. Chapter 6.4 will demonstrated how the service level can be considered in a passenger car fleet model.

3.7 The Rebound Effect in passenger car transport

The basic idea of efficiency improvement measures is the reduction of energy consumption. From a pure technical perspective the energy consumption should be reduced to the same extent as the efficiency is improved by the implemented measure. However, in practice the energy savings are usually smaller, due to a range of mechanisms that are summed up as "rebound effects" (cf. (Sorrell 2009)).

Improvements in efficiency usually lead to reduction in the cost of goods or services. According to the economic demand theory this leads to an increased demand that reduces the savings that have been achieved. In passenger car transport for example, efficiency improvements of cars reduces fuel cost per kilometre. This cost reduction effectuates that users drive more and thereby offset a certain part of the energy savings (see Figure 3-3). This travel rebound has been proved by (Schipper et al. 2002) who showed that the lower fuel costs of diesel cars in Europe in the 1990ies have caused considerably higher yearly driving distances which have offset part of the fuel savings.

Furthermore, the increased demand for this transport mode will also be reflected in the fleet size. Lower service costs attract more consumers to buy a car which causes further energy consumption.

Another rebound effect is the increase of service level. Consumers tend to use efficiency improvements in services to increase their service level. In the case of passenger car transport this means that they buy bigger and more comfortable cars which offsets some of the savings the more efficient technology would provide if applied in a vehicle with the same service level. This inter-relation is approved by historic analyses that have also showed that the increasing service level in the passenger car fleet has often outweighed efficiency improvements achieved through technological progress (Van den Brink & Van Wee 2001) (Zervas & Lazarou 2008) (MIT 2008).

Apart from the above mentioned direct rebound effects there are also indirect rebound effects. The cost savings resulting from more the higher efficiency of the transport service are often used to consume other services that also require energy.

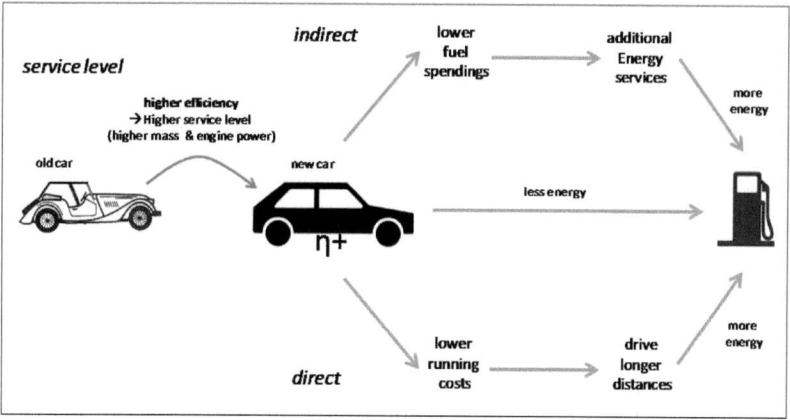

Figure 3-3 : Rebound effects in passenger car transport (adapted from (Sorrell 2009))

It is obvious that direct rebound effects have to be considered when modelling the impact of changing framework on energy consumption of the passenger car fleet. In chapter 6 it will be presented how direct rebound effects can be captured in an energy economic model of passenger car transport.

3.8   Costs of Transport

As indicated in the previous sections costs are an important factor for economic analyses in the transport sector. When talking about costs in an energy economic context it is important to differentiate between internal and external costs:

Internal costs or private costs of transport are the costs that have to be borne by the consumer of the transport service. This includes fixed costs of the transport mean and all running costs, like energy costs, maintenance costs, road charges etc (see (Maibach et al. 2008)). A detail overview on internal costs of passenger car transport in Austria will be given in chapter 5.8.

External costs or social costs include all costs that are not directly borne by the consumer of the transport service. This includes costs for infrastructure, environmental costs, congestion costs, accident costs etc.

The internalisation of these external costs is one major problem in the transport sector. If external costs are not internalised in the transport cost, consumer decisions are influenced in a way that might lead to welfare losses (Essen et al. 2008). Therefore, a correct internalisation of external costs is an important issue in the transport sector and can be an effective instrument to reduce negative side effects of transport.

According to (Essen et al. 2008) "*it may:*

- Improve economic and in particular transport efficiency (e.g. efficient use of energy and of scarce infrastructure and rolling stock of all transport modes).
- Guarantee a level playing field between transport modes
- Improve safety and reduce environmental impacts of the transport sector."

It is the task of policy makers to create framework conditions that assure a correct internalisation of external costs in the transport sector. In practice there are two major ways to internalise external costs of transport. It can be done directly through regulatory measures or indirectly through market based instruments like taxes, charges or other instruments (Maibach et al. 2008). A detailed overview on policy strategies in the passenger car sector in Austrian and other European countries are given in chapter 2.6 and 4.6.4.

3.8.1 Internal Cost of passenger car transport

For an economic assessment of this transport mode the total cost of ownership (TCO) has to be considered. In the case of a passenger car this includes fixed costs, like the capital costs of the vehicle, insurance costs and tax on ownership, as well as variable costs, like fuel costs, operational costs (including car maintenance, lubricants, tyres etc.) and costs that arise from the use of the infrastructure.

$$TCO = CC + FC + OC + INS + TO + CI \quad \text{[€ year}^{-1}\text{]} \quad (3\text{-}10)$$

TCO ... total cost of ownership [€ year^{-1}]
CC ... capital costs for the car [€ year^{-1}]
FC ... fuel costs [€ year^{-1}]
OC ... non fuel operational costs [€ year^{-1}]
INS ... insurance costs [€ year^{-1}]
TO ... tax on ownership [€ year^{-1}]
CI ... cost for use of infrastructure (parking and road charges) [€ year^{-1}]

The most important type of cost in the case of a passenger car is usually the capital costs for the car. The capital costs include the net costs of the vehicle, tax on ownership (if applicable) and value added tax.

$$CC = CC_{net} + TA + VAT \quad \text{[€]} \quad (3\text{-}11)$$

CC... initial gross capital costs [€]
CC$_{net}$...net capital costs [€]
TA ... tax on acquisition [€]
VAT ... value added tax [€]

For a correct economic combination of initial capital costs and the other types of costs that are occurring during the vehicle life time, the annuity of the capital costs is calculated considering the discount rate and the depreciation time of the car by using the capital recovery factor.

$$CC_{SP} = (CRF \cdot (CC + TA + VAT)) \quad \text{[€ year}^{-1}\text{]} \tag{3-12}$$

$$CRF = \frac{r \cdot (1+r)^{DT}}{(1+r)^{DT} - 1} \tag{3-13}$$

CRF ... capital recovery factor
r ... interest rate [%]
DT ... depreciation time [years]

The fuel costs are usually the second most important cost type for passenger cars. The main factors affecting the fuel costs are the fuel price, the energy consumption of the car and the distance driven. Apart from the net energy prices fuel costs usually include, fuel taxes and the value added tax.

$$FC = FP \cdot EC \cdot D \quad \text{[€]} \tag{3-14}$$

$$FP = (FP_{net} + FT) \cdot VAT \quad \text{[€ litre}^{-1}\text{]} \tag{3-15}$$

FC ... fuel costs [€ km^{-1}]
FP ... gross fuel price [€ litre^{-1}]
FP$_{net}$ ... net fuel price [€ litre^{-1}]
FT ... fuel tax [€ litre^{-1}]
EC ... energy consumption of the car [l 100km^{-1}]
D ... distance driven [km]

3.8.2 Specific service costs

To fit the passenger car transport costs in an energy economic framework, the total cost of ownership are broken down to the specific service costs of the transport mode expressed in cost per kilometre.

$$SC = \frac{TCO}{D} \quad \text{[€ km}^{-1}\text{]} \tag{3-16}$$

SC ... service costs of passenger car transport [€ km^{-1}]

As indicated in the beginning of this chapter the specific service costs are a major economic factor for all transport modes and will play an important role in the energy model presented in chapter 6.

3.9 Transport demand and service level in passenger car transport in Austria

As indicated in chapter 2.1 transport-related energy consumption has increased considerably in the last decades in Austria. The main driver of this development has been the growing car fleet (see Figure 2-4). While the fleet has grown considerably between 1990 and 2008 the average kilometrage of cars remained more or less constant (see Figure 2-11 in chapter 2.5).

As mentioned above the energy consumption of the transport sector is determined by the transport demand and the efficiency respectively the service level of the cars. These parameters are mainly affected by the income and the cost of transport. The following chapters will give an overview on the historic developments of these parameters in Austrian and their impact on the passenger car fleet.

3.9.1.1 Income

As described above income is the most important driver of transport related energy consumption. It is not only affecting car ownership and car use but also the service level of the cars sold.

Also in Austria the real GDP development between 1990 and 2008 (see Figure 3-4) was a strong driver for the increasing energy demand of passenger car transport. It led to a strong increase in car ownership and consequently to higher cumulative transport kilometres in the fleet.

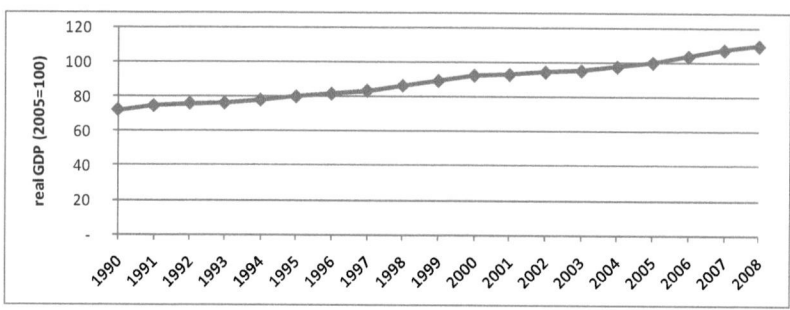

Figure 3-4: Real (2005) GDP development in Austria (Data Source: (Statistics Austria 2010c))

3.9.1.2 Fuel cost

Real fuel cost of passenger car transport is determined by real fuel price and the specific fuel consumption of cars. Between 1990 and 2008 there were ups and downs in real fuel price development but considering the whole period the real fuel price remained more or less constant (see Figure 3-5). Only the diesel price increased considerably between 2003 and 2008 mainly driven by the growing demand generated by the increasing number of diesel cars in the European passenger car fleets (ACEA 2010).

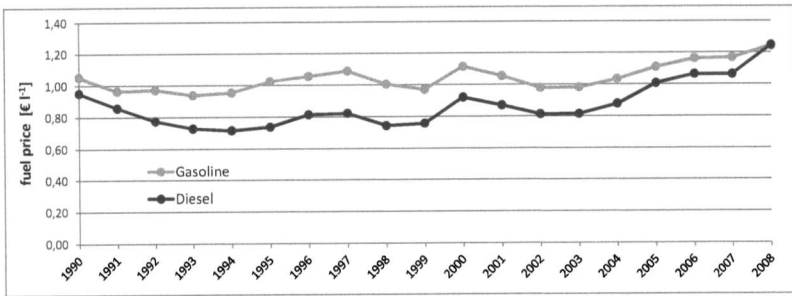

Figure 3-5: Real (2008) fuel price in Austria 1990-2008 (Data Sources: (Fachverband Mineralölindustrie 2010b) & (Statistics Austria 2010a)

There is a significant difference between specific fuel costs of gasoline and diesel cars in Austria. Fuel costs of diesel cars have been 20 % to 30 % lower compared to gasoline cars between 1990 and 2000, due to their better efficiency and the lower tax on diesel. This cost gap has driven many consumers to buy diesel cars and has led to a strong increase in the share of diesel cars in the fleet. The development is also reflected in the average fuel cost of cars in Austria which decreased significantly between 1990 and 2003 (see Figure 3-6). This real price decrease of specific fuel cost is one explanation for the growth in demand and service level in the passenger car fleet in this time. Between 2003 and 2008 the price increase of both diesel and gasoline fuels lead to a real increase in fuel cost.

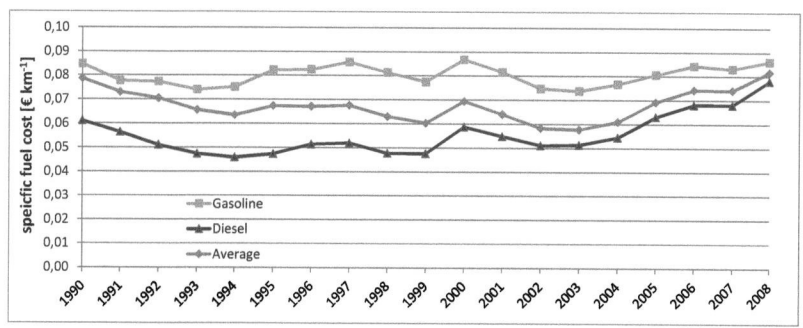

Figure 3-6: Real (2008) specific fuel costs of gasoline and diesel cars in Austria 1990-2008 (Data Sources: (Fachverband Mineralölindustrie 2010b), (Odyssee 2010))

4 Technological change in propulsion technologies for passenger cars – A historical survey

Today the internal combustion engine (ICE) is the standard propulsion technology for passenger cars. It serves as the central energy conversion unit in conventional and even in most alternative powertrain systems. The main arguments that have led to the dominant position of ICE based systems can be traced back to the energy carriers they are using. The standard fuels of internal combustion engines today are either gasoline or diesel, both mainly based on fossil crude oil. With their high energy density and their excellent storage capability, in liquid aggregate state at atmospheric pressure and at a broad temperature range, they are the optimal energy carrier for mobile application like motor vehicles. The combination of these fuels and internal combustion engines turned the motor vehicle into an attractive transport mean for both freight and passenger transport and initiated a new era of transport and mobility.

In the beginning of the 20th century the number of motor vehicles in industrialized countries increased steeply and they replaced horses as mean for passenger transport within a relatively short time period (Nakicenovic 1986) (see Figure 4-1). This development points out with which dynamics technological changes can evolve when the new technology offers essential advantages compared to an old one.

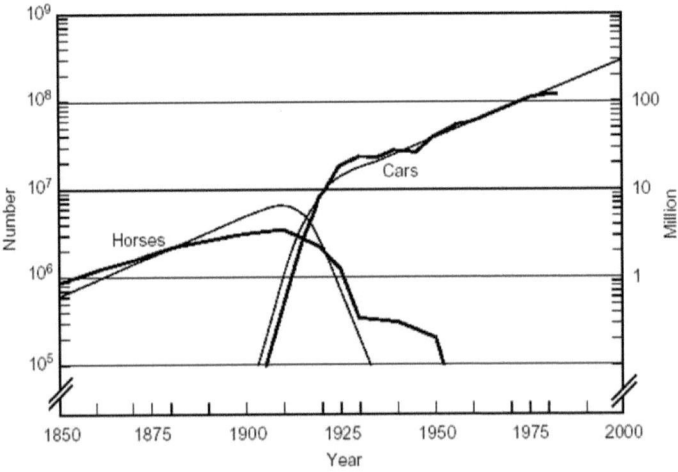

Figure 4-1: Number of (urban) draft animals (horses) and automobiles in the USA, empirical data (bold jagged lines) and estimates (thin smooth lines) from a logistic model of technological substitution. (Nakicenovic 1986)

4.1 Technological Life Cycles and Technological Diffusion

Generally the development of technologies is defined by three different phases: invention, innovation and diffusion (Grübler 1998). The invention represent the first prove of technical feasibility of a new solution. An innovation marks the point when an invention is used in a practical application for the first time. The phase when an innovation becomes applied on larger scale and gains market share is called the diffusion of an innovation. Diffusion processes of technological innovations follow a characteristic S-shaped scheme and can be represented by technology life cycle models that cover the characteristic phases of the development of a technology. The main phases are the *Introduction* or *childhood* phase, the *growth* phase and the saturation or *maturity phase* (Grübler 1998).

The validity of this theory has been proved in many technological diffusion processes, e.g. transportation infrastructures (Nakicenovic 1991), steel production techniques (Nakicenovic 1987b), marine propulsion technologies (Nakicenovic 1987a).

The diffusion process can also be split up in phases named after the groups that are adopting the technology in those phases: Innovators, Early Adopters, Early Majority, Late Majority, Laggards. Those adopter groups and the characteristic shape of the diffusion curve were first described by Everett Rogers in 1962 (Rogers 2003).

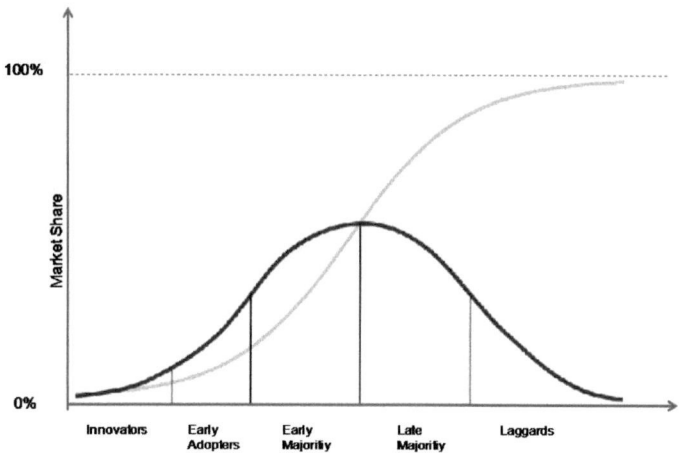

Figure 4-2: Rogers Curve (adapted from (Rogers 2003))

4.2 Diffusion of technologies for passenger cars

In the future efficiency requirements of passenger cars will ask for new, alternative propulsion technologies. Today these technologies are only partly available and most of them are in an early status of development. However, for an assessment of future developments in the passenger car transport future technology options have to be considered. For long term scenarios it is important to know the necessary time frames and the conditions for technological changes.

To model the dynamic of potential future diffusion processes that might affect the passenger car sector in the next decades, past technological developments have been analysed. In automobile history there have always been technological innovations that have diffused into the car market (see Figure 4-3). Some of these innovations were additional features that improved the cars comfort (e.g. air conditions) or safety. However, most of these innovations have simply replaced old technologies. (Nakicenovic 1986) describes such diffusion processes of automotive technologies giving a comprehensive insight in their dynamics. The analysis includes among others innovations in transmission, breaking and exhaust gas after treatment systems. A more recent analysis focuses on innovations that improve the motor efficiency, like downsizing (Cuenot 2009) (see figure Figure 4-4).

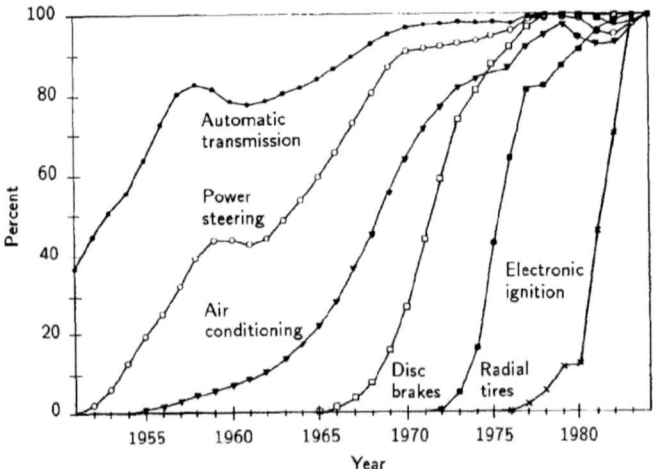

Figure 4-3: Diffusion of technologies in the US car industry, percentage of car manufactured (Source: (S. Jutila & J. Jutila 1986))

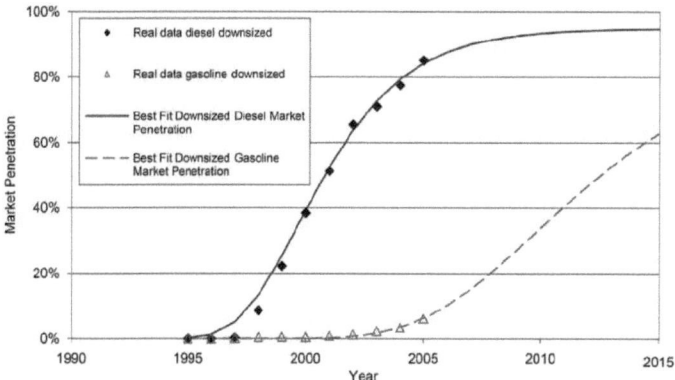

Figure 4-4: Share of downsized engines in European passenger car sales (Source: (Cuenot 2009))

The problem with these past diffusion processes in propulsion technologies is the fact that they all rather represent technological evolution steps than revolutions. In fact in the last century of automobile use there has not been a real technological revolution in the propulsion of the car. For over a hundred years the propulsion system of motor vehicles was based on internal combustion engines running on liquid hydrocarbon fuels. There were many innovations that have improved this technology considerably (Nemry et al. 2008), but none of them can be considered a revolution.

Revolution developments in the field of propulsion systems would mean the shift toward a completely different technological principle. Switching from conventional cars to battery electric cars or hydrogen based fuel cell cars, can be seen as revolutionary steps. Such steps implicate radical changes in the entire field of passenger cars, including vehicle users, car industry, fuel suppliers, infrastructure etc.

When modelling of future scenarios it is important to consider the radicalness of such innovations. In chapter 6 it will be presented how the technological diffusion theory can be applied for modelling future technological diffusion processes in the field of car propulsion technologies.

4.3 Technological Learning

The main drivers of the diffusion process of an innovation are the so called learning effects. When entering the market an innovative technology is usually more costly than the established and mature technologies. Later on, the cost of the technology decreases as consequence of increasing production experience (learning by doing) and higher production scales (economics of scale). These two effects, learning by doing and economics of scale, are both considered in the theory of technological learning.

According to this theory the cost of a technology can be seen as function of its global cumulative production and its learning index (see equation 4-1) (cf. (Grübler 1998)). From the learning index b the progress ratio q can be derived. The progress ratio expresses what cost reduction are effected by a doubling of cumulative production (percentage of former costs after a doubling of production).

$$C(x) = a \cdot x^{-b} \qquad (4\text{-}1)$$

$$q = 2^{-b} \qquad (4\text{-}2)$$

C ... cost per unit [EUR/unit]
a ... cost of first unit produced [EUR/unit]
x ... number of produced units
b ... learning index
q ... progress ratio

Another term that is often used to express learning effects is the so called Learning Rate LR, which also indicates the cost reduction of the technology:

$$LR = 1 - 2^{-b} \qquad (4\text{-}3)$$

LR ... learning Rate

Since the estimated cost development depends on the cumulative production and the learning index the estimation of these two parameters always implicates uncertainties in this method.

A possible approach would be to use learning parameters derived from past technologic developments. The range of learning rates for energy related technologies thereby extends from 5 % to 25 %, with an average of around 16-17 % (McDonald & Schrattenholzer 2001).

The other critical parameter is the global cumulative production. The obvious solution for this problem would be to use endogen learning effects within the model. However, most models are limited to a specific country or region, which means that endogenous learning effects cannot be applied as they are referring to a global dimension.

In this thesis technological learning is applied to estimate the future cost of key components of future propulsion systems for passenger cars. The chosen approach will be explained in detail in chapter 5.6.

4.4 Diffusion Barriers

Once a diffusion process of an innovation is initiated it is not guaranteed that the innovation will actually succeed and become the technological standard. Even though the innovation might have advantages in some fields there are also factors that can avoid or slow down the diffusion process. These so called diffusion barriers can have very different reasons and characteristics.

In the case of passenger car propulsion technologies one main barrier might be the cost of the technology. New technologies usually start in smaller production scales and they cannot benefit from the production experiences of thousands or even millions of past units produced. That is why they are usually more costly in the beginning. However, when an innovation has evident advantages for the consumer the demand for it increases and therewith cost decreases as a consequence of learning effects. In the modelling of technological diffusion processes the cost barrier is usually captured through learning effects (see chapter 4.3).

There are also other barriers that might slow down the diffusion process of a technology. In the case of propulsion technologies for passenger cars these barriers could have different reasons. For example if the technologies require other fuels than the ones used by conventional technologies, the lacking availability of refuelling infrastructure can be a serious barrier (see (Aral 2009)). This barrier is also the main reason why CNG cars are still a niche technology in Austria today even though they are a cost effective alternative (Kloess et al. 2009). For hydrogen or electricity bases technologies the infrastructure might become an even higher barrier.

Another diffusion barrier can be the availability of the technology for example when the range of car models available with the desired propulsion technology is very small which could drive consumers to choose other options. When the consumer intents to buy a car he can chose between hundreds of models from different producers. However, when he prefers a special alternative propulsion technology the choice is often sharply limited. What has to be considered is that the consumer decision for buying a car is strongly influenced by practical and emotional aspects of the car. The propulsion technology mostly represents a secondary criterion. Therefore, it is not a primary concern for the car producers to offer every propulsion technology for every model, especially when an additional variation of a model causes high extra cost in the production process. Only if the consumer really starts to ask for the technology the producers would react and bring the demanded models on the market. A perfect example for this is the diffusion of turbo charged diesel engines in passenger cars in Europe starting in the 1990ies (see chapter 2.3 & 2.4). At the beginning there were only few producers that had these models in their portfolio. The great success of these models forced all competitors to follow until almost every model was available with both, gasoline or diesel engines. This process took about ten to fifteen years and gives an impression how these types of barriers can slow down technological diffusion in this field.

Another diffusion limitation is the simple fact that the majority of consumers tend to be conservative in their decisions, especially when it comes to large investments as it is required when buying a new car (Aral 2009). This means that they would rather buy technologies, proven to be reliable and efficient in the past. That is why new technologies are only slowly increasing their market share even after becoming cost efficient.

4.4.1 Lock-In phenomenon

The so called Lock-In phenomenon is often one of the greatest barriers in the technological diffusion process. Thereby, the learning effects of an established technology make it difficult for new technologies to compete even if they have technologic advantages and could become competitive. The established technologies benefits from significantly lower cost due to past learning effects (G. Erdmann & Zweifel 2008).

In (Cowan & Hultén 1996) it is described how the automobile finds itself locked-into the use of internal combustion engines and hydrocarbon as primary propulsion technology. It turned out to be very difficult to escape this situation because of various reasons: Firstly there are past learning effects as described above. Secondly producers would rather prefer to keep oneselling the established and depreciated technology than investing in a new alternative (G. Erdmann & Zweifel 2008). Thirdly *"consumers are unwilling to switch technologies because they have invested time and money in the technology that dominates"* (Cowan & Hultén 1996).

These factors will always make it difficult for any alternative propulsion technology to diffuse into the market especially when it incorporates revolutionary changes like electric or fuel cell cars.

4.5 Historic technological trends in passenger car propulsion systems

Today passenger cars almost exclusively run on internally combustion engines, namely piston engines fired by liquid hydrocarbons. However, there have always been alternative propulsion technologies. Especially in the early years of motor vehicle development in the beginning of the 20th century it was not obvious which option would be most adequate to be used for motor vehicles. During the over 100 years of motor vehicle history there have been many attempts to introduce new alternative technologies, but they so far never succeeded.

The following chapters will give a short review on technological trends in propulsion technologies for passenger cars covering the early development of motor vehicles that lead to the breakthrough of the piston engine. Furthermore, an overview on recent attempts to introduce alternative vehicle technologies will be given.

4.5.1 The breakthrough of the internal combustion engine

In the early years of the motor vehicle history internal combustion engine based cars were seriously challenged by electric vehicles and also by steam cars. At the turn of the 19th to the 20th century the three technologies were about head-to-head in terms of yearly sales (Cowan & Hultén 1996). In this early stage the EV was simply the more attractive technology. While the ICE cars were loud and emitted malodorous exhaust gases, EVs were clean and silent and they needed no mechanic starter for the engine (Naunin 1994). Therefore, EVs were the first choice for the early users of motor vehicles, who mainly came from upper social classes and used the car as a status symbol. Also the limited driving range was not seen as serious problem back then. It has to be considered that the road infrastructure we have today is a result of the steady motorization process during the twentieth century and did not exist in the pioneer days (see Figure 4-5).

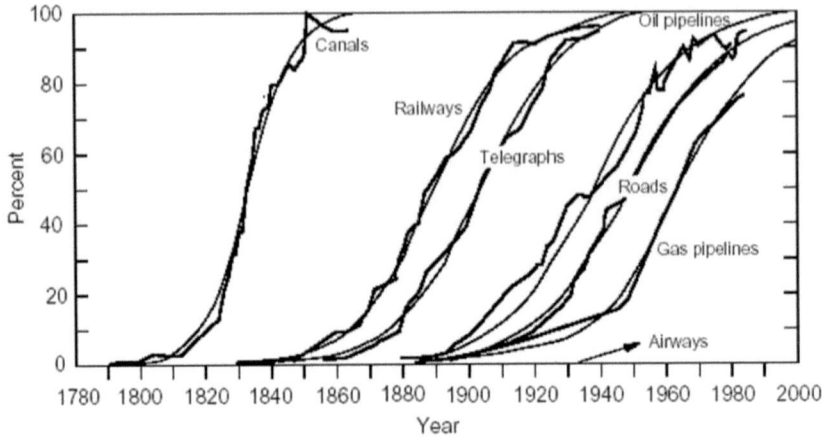

Figure 4-5: Growth of US transport infrastructures as a percentage of their maximum network size, empirical data (bold jagged lines) and model approximation (thin smooth lines). (Source: (Nakicenovic 1987b))

One of the key developments that lead to the breakthrough of the ICE as standard propulsion technology was the electric starter that made the technology much more attractive to consumers (Naunin 1994). With the development of the road infrastructure and a dense refueling infrastructure for ICE-cars the low driving range became a clear disadvantage for EVs that finally led to their disappearance.

Once the ICE car was established as standard it also defined what consumers expect from a passenger car in terms of driving performance and comfort level. This and the fact that the technology has been continuously improved with strong effort on a global level, made it difficult for every alternative technology to compete. Till this day there has been no technology that could seriously challenge the internal combustion engine as standard propulsion technology for motor vehicles.

4.5.2 Electric Mobility Hype (1990)

In the early 1990ies there was strong development boost for electric cars triggered by the implementation of the Clean Air Act in California in 1990. The clean Air act dictated that 2 % of cars registered have to be zero emission vehicles (ZEVs) in 1998 and 10 % in 2003 (Strock 1996). This clean air act created enormous pressure on car makers to develop zero emission cars and led to strong research and development efforts by most major car companies. A consequent implementation of this law would have meant that all car companies that wanted to sell vehicles in states where the law was implemented would have to sell a certain percentage of real zero emission cars. Furthermore, the law could have set a standard for other industrialized countries and thereby could have affected major parts of the global car market.

To face this serious threat, carmakers put high effort on the development of zero emission cars. The best technological option hereby was the electric car. The basic technology of electric cars was well known and available and there also had been past experiences from prototypes. Consequently, the development efforts of the car industry led to the development of several prototypes and even some series cars after just a few years in the 1990ies. Most cars were just conversions of existing conventional models, but there were also models designed especially as electric cars (e.g. GM EV1). Furthermore, there were public support programs for electric vehicles (Christl et al. 1992) and public funded fleet testing programs (e.g. (Voy 1992)) .

However, during the 1990ies the law was more and more diluted due to intervention of lobbying groups. Finally, the definition of zero emission vehicles ZEV was adapted in a way that even conventional cars with combustion engines and sophisticated exhaust gas after treatment could meet its criteria. This made the development of real ZEVs dispensable and caused an abrupt end of all efforts of car makers to develop electric cars. After that most existing vehicles were withdrawn from the fleets immediately.

4.5.3 Hydrogen & Fuel Cell Hype (2000)

The clean air act in California in the 1990ies was also a strong driver for another technology that was aimed to be the successor of the internal combustion engine. In the 1990ies car makers made strong effort to develop hydrogen based cars. These cars were seen as the long term future for an emission free and carbon neutral passenger transport.

There are mainly two options of using hydrogen as a motor fuel. One is to burn it in conventional internal combustion engines, an option that was followed by BMW and Ford (Kloess 2006). The other and more ambitious approach are fuel cell cars. In terms of efficiency fuel cells are superior to ICEs (see chapter 5.3), but they are also much more challenging from a technological perspective. Nevertheless, the major car companies, above all Daimler Chrysler, GM, Toyota and Honda, have invested in this technology with the ambitious goal of a short to medium term introduction. In addition to the activities of carmakers there was strong financial support by public authorities with public funded technology development- and fleet testing programs (Begluk 2009).

All stakeholders were very confident that the technical progress would lead to a mass market introduction of hydrogen based cars soon. For example in the late

1990ies Daimler Chrysler dated the mass market introduction of fuel cell cars to 2004, and five year later this date was postponed to 2010 (Weider et al. 2004).

The main reasons why this ambitious plans have failed is the fact that it has been impossible to bring fuel cell costs down to a level where a large scale market introduction was feasible. Apart from the fuel cell itself the hydrogen storage is still a problem that has not been solved satisfactory. Another unsolved problem is the hydrogen refueling infrastructure. Hydrogen based cars thereby have to face a classical chicken-egg problem that could not be overcome in a short term.

4.5.4 Electric Mobility Hype (2010)

Between 2005 and 2010 it became obvious that fuel cell cars would not become feasible in a short to mid-term. This drew attention back on battery electric vehicles, a technology that appears more mature and closer to potential commercialization. This time it weren't the big car companies who triggered this hype, but small companies who demonstrated that it is possible to build roadworthy electric cars even with low budgets (e.g. Tesla Roadster). Furthermore, the progress in battery development significantly improved the performance characteristics of EVs and made them much more attractive and promising than back in the 1990ies. In the view of this growing hype the car industries were forced to draw more attention to this topic. However, at this time the official long term technology vision of most car makers was still the fuel cell car (Kloess 2006).

In the years 2007-2010 more and more prototypes of EVs and plug-In hybrid vehicles (PHEVs) were presented at motor shows and the mass market introduction of pure electric cars is announced for the following years.

In the last years EVs also became subject to enforced public funding. Very similar to the support for fuel cell cars ten years before and for electric cars around 1990, there is large public funding of research programs and fleet tests.

EVs have created strong ambitions even in other industry sectors. Utilities see them as future electricity consumers and get strongly involved with research and fleet testing programs.

Summing up there are clear parallels to the activities that were seen around 1990. There are many prototypes, very few EVs on the market, public support programs for the cars and for R&D and intensive fleet testing programs. This could raise the question why EVs should succeed this time.

One main difference between the 1990ies is the real price level of fossil fuels. The high price of the last years has been important driver of efficient cars. Another key factor will be the status of electricity storage technologies. Furthermore, today there are technologies that could help the transition toward pure electric cars (e.g. hybrid cars HEV, plug-In hybrid cars PHEV). (see chapter 5.2)

4.5.5 Key findings from past technology "hypes"

Comparing the past two hypes for alternative propulsion technologies, the hype on EVs in the early 1990ies and the hydrogen and FCV hype around the millennium there are many parallels to the present hype on EVs and PHEVs: On the peak of all three hypes the perception was created that we were at the beginning of a new era of automotive propulsion technology and in 10-20 years passenger car transport would be fundamentally different. However, nothing of these visions became true and today in 2010 the internal combustion engines is still the dominating technology worldwide.

In retrospect it is remarkable how the technological development has been overestimated. In particular in the case of fuel cell cars the cost reduction projections were far too ambitious. Up to now the high investment in fuel cell vehicles has been an economic disaster and it is uncertain if any of these investments will ever lead to a profitable product.

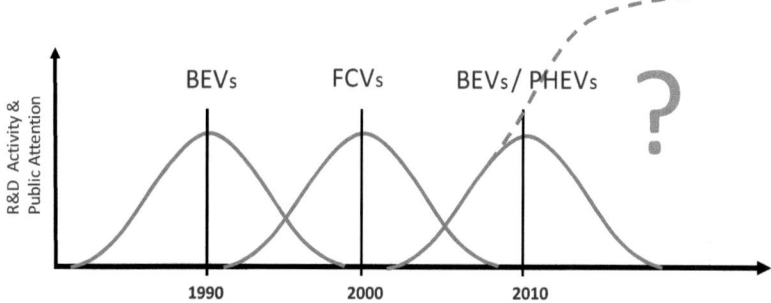

Figure 4-6: Technological "Hypes" in alternative vehicle propulsion technologies

In the future automobile producers will have this fuel cell vehicle disaster in the back of their minds when it comes to investing in other revolutionary technologies. As a consequence the car makers' strategy in the present hype for EVs and PHEVs rather reacting instead of acting. This could be a possible interpretation why there has been no hurry in being the first to bring an EV on the market and why there is currently only one EV model of a major car maker on the market (Mitsubishi I-miev (Mitsubishi Motor Corporation 2010)).

Contrary to the 1990ies today there is no political framework that could force the car industry to build these cars. Today, public authorities limit themselves to promoting EVs and PHEVs rather than implementing compulsory regulations like in the early 1990ies. Summing up it is doubtful whether support instruments and environmentally aware consumers will be enough to turn the present hype into a real technological diffusion process.

All these facts have to be kept in mind when estimating the technology mix for future scenarios for passenger car transport.

4.6 Stakeholders for the diffusion of vehicle propulsion technologies

As described in chapter 2 the development of the passenger car sector in terms of quantities (fleet size and kilometrage) is determined by parameters like income, price of the transport mode, infrastructure etc. The development of these parameters is strongly influenced by the stakeholders in this sector. The main stakeholders of passenger car transport are political authorities, the car industry, the fuel industry and above all the consumers. In the following chapters the role of each of these stakeholders will be analysed to assess their interests and motives concerning the future development of passenger car transport in terms of transport volumes, costs, efficiency and technologies.

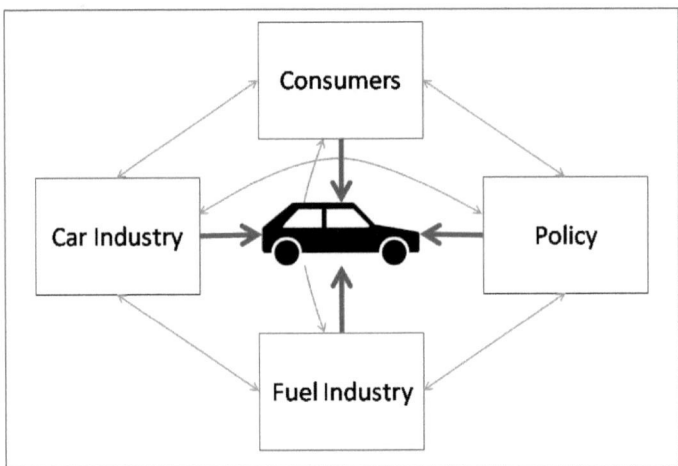

Figure 4-7: Main stakeholder for passenger car transport

4.6.1 Consumers

In Austria the passenger car is the most important mean of passenger transport and plays a main role in the entire mobility pattern of the country (see chapter 2). The dynamic motor vehicle diffusion in the 20th century has massively shaped the countries social and economic development. Today the functioning of major parts of the country's economy is based on the use of motor vehicles.

With the widespread use of passenger cars individuals got much higher regional flexibility, with many advantages for both the industry and the employees.

As described in (Aberle 2003) the average time of the day consumers spend for transportation has remained constant the last decades (e.g. 60 – 70 min in Germany). However, with the improvement of infrastructure and the availability of new means of transport, like the passenger car, the distance that could be travelled in this time increased constantly. In developed countries passenger cars are available for large parts of the population today and the infrastructure and the economy are to a high extent oriented toward this transport mode.

The advantages of this high degree of individual mobility and flexibility are evident. However, there are also serious downsides of this development. Growing road congestions as a consequence of continuously growing traffic volumes and increasing emissions of pollutants and greenhouse gases are only some examples. Furthermore, the growing dependence to passenger cars has created other serious problems, which are not that evident to most consumers. In the past decades the society and the economy has become increasingly dependent to the use of motor vehicles, partly creating a *lock-in* in this mode of transport (see chapter 4.4.1). The dominance of the passenger car is so overwhelming that more and more transport alternatives are continuously repressed. This development could be observed in several industrialised countries where the relative importance of passenger car transport decreased considerably (see (Button 2010)). This transport mode *lock-in* together with the *lock-in* to internal combustion engine-based propulsion system has created a high dependence to this transport mode and the hydrocarbon fuels it is based on. This makes consumers economically vulnerable to oil price fluctuations, an aspect that is often neglected in the consumers' decision making process.

In many countries and regions this dependence is so advanced that consumers often have no chance to escape this *lock-in* situation as the exit strategy would simply be too costly (e.g. higher prices of real estate in city centres or with good access to public transport).

Alternative propulsion technologies can be an option to at least partly escape this *lock-in* situation. However, every alternative technology has to struggle with the side effects of the current lock in situation. Firstly there is the economic dimension of the lock in effect described in chapter 4.4.1 that makes it difficult for alternative technologies to compete. Secondly there are the other diffusion barriers (e.g. lack of refuelling infrastructure and lower driving range) that make alternative technologies less attractive to consumers (see chapter 4.4). Thirdly the consumers' expectations of passenger cars are defined by the characteristics of conventional cars: extremely high driving range of up to 1000 km without refuelling, the possibility to refuel at every petrol station within a few minutes, high comfort level etc. These expectations are difficult to meet by alternative options (Aral 2009).

An escape from this lock-in situation can only be achieved through a combination of alternative technologies with appropriate economic and political framework conditions.

4.6.2 Car Industry

The car industry has been one of the great profiteers of the trend toward passenger car transport. In many industrialized countries the car industry represents one of the most important industry branches with thousands of jobs directly or indirectly dependent to them. This gives them a very strong negotiating position when it comes to enforcing their interests with public authorities. It is evident that the car industry as a stakeholder cannot be interested in a reduction of passenger car transport and therefore they have to be a clear opponent of all developments and tendencies that threaten their core business, due to simple economic interests.

One serious threat to their business could be high fossil fuel prices. An increase in fossil fuel price could lead to significant increase of real transport cost and thereby reduce the demand for passenger cars (see chapter 3.2). The car industry tries to prepare for this threat by developing fuel efficient cars which help to keep real transport cost low and demand for cars stable. Concerning propulsion technologies the industry is still very focused on the internal combustion engine (ICE). Even most (so called) alternative propulsion technologies they offer are still based on ICEs (e.g. CNG cars and hybrid cars). From a pure economic standpoint it is evident why they have to focus on this technology. As described in chapter 4.6.1 consumers' expectations of a car are strongly biased by the characteristics offered by conventional ICE based cars. Car manufacturers would only offer alternative technologies that can meet these expectations. Electric cars with their limited driving range and lower comfort levels are not really an attractive product with high market potential. From an economic perspective it simply makes no sense for the car industry to invest billions in the development of cars that are very likely to be a commercial flop.

Another possible argument against the introduction of electric cars is the simple fact that the introduction of this technology could in some day make ICE technology obsolete. This would mean that all the money that was invested in this technology would turn into *sunk costs* a threat that firms in all fields have to fear when it comes to radical innovations (see (Satorius & Zundel 2005)). The established players in the automotive industry are all in the same situation and no one of them would move towards such technological revolutions unless it is not necessary to remain competitive. Once one player risks the step and offers

EVs in a large scale, and is successful, all competitors will have to follow. But as long as this does not happen the dominant strategy of a car manufacturer has to be to wait and see.

That is pretty much the situation we have now (2010). All car makers show that they are capable to build electric cars, by presenting prototypes at motor shows or by taking part at fleet testing programs, but most of them would rather prefer not to be the first who takes the risk of bringing it to the large scale market. Hereby, the potential high risk overweighs the incentive of being perceived as the innovator on this field, and therefore profit from so called *first mover advantages* (see (G. Erdmann & Zweifel 2008)).

Another potential implication of the emergence of alternative propulsion technologies like electric vehicles is the innovative pressure they create. As indicated above their efficiency is outstanding in comparison with the ICE-based cars today. However, ICE based cars still have considerable potential of efficiency improvements. Therefore, alternative technologies might create innovative pressure that could boost the efficiency of conventional cars by enforced application of efficiency improving measures (e.g. weight reduction, drag reduction, motor downsizing, hybridisation etc.)

Altogether it can be concluded that the car industry won't be a strong driver for electric cars.

4.6.3 Oil Industry

Another important stakeholder in the diffusion of alternative vehicle propulsion technologies is the oil industry. As provider of the fuel and the corresponding infrastructure the oil industry plays leading part in the passenger car sector. The sale of motor vehicle fuels is their core business and passenger cars are the biggest consumers of these fuels. Most oil companies are to some extend vertically integrated and principally focused on crude oil products and to some extend on natural gas. This means that their business segment does not only include the sale of the fuels to the consumer but often the entire supply chain including oil production and refining. In the refinery process different products are extracted from crude oil that altogether contribute to the economic profit of the Industry. The major part of the crude oil is used to produce different types of transport fuels, like middle distillates (mainly diesel fuel), gasoline, heavy oil and aviation fuels (IEA 2009a).

Since the oil industry strongly depends on the sale of transportation fuels they will try to keep up demand for these products. Alternative fuels are only acceptable when thy fit into their fuel supply chains and into their business model, that relies not only on the fuels sold but also on the turnover generated in the shops of fuel stations.

Hence, it is evident that the diffusion of electricity based propulsion technologies represents the worst case scenario for this industry, since electric cars would make diesel and gasoline obsolete and leaf their refuelling stations (and shops) abandoned.

Consequently the oil industry has to strongly oppose the diffusion of electric cars.

4.6.4 Policy

The fourth and maybe most important stakeholder for developments in the field of passenger cars is the public authority setting the political framework for this mean of transport. As a stakeholder in the diffusion process of efficient or alternative propulsion technologies the public authority is the only player who could be interested in a shift toward environmentally more benign technologies or fuels. If transport policy follows the concept of internalisation of external costs consequently this would inevitably lead to promotion of efficient and clean technologies (cf. chapter 2.6 & 3.8). However, in practice internalisation of external costs is not the only objective function for policy. There are many other factors and strong interest groups that have to be considered by policy makers. The opponents of higher motor vehicle taxation usually argue that higher taxes can threaten the country's economic competitiveness and the resulting price increases can be an economic burden to people who are dependent on the use of passenger cars (e.g. commuters). Furthermore, the car industry itself also has considerable influence on policy makers especially in countries with many jobs involved in this branch (e.g. Germany, USA). That is why the political framework will always be a compromise between the basic goals and the involved interest groups. This means that radical changes are very unlikely to happen in a short term.

The strong impact of policy as a stakeholder in passenger transport can be described on the example of Denmark. As illustrated in chapter 3.4 car ownership in Denmark is relatively low when taking in account the country's high income level. A main reason for this apparent contradiction can be found in the high taxes in this country, with very high taxes on fuels and the highest taxes on vehicles in the European Union (Ajanovic et al. 2009).

5 Techno-economic assessment of hybrid and electric propulsion technologies for passenger cars

Increasing fossil fuel prices and emission reduction commitments will ask for higher efficiency and lower emissions in future passenger cars. Electrification/hybridisation of the vehicle powertrain is an important approach to improve overall efficiency of passenger cars and is often seen as a first step toward electric mobility.

The following chapter gives an overview on this development. In a first step it will briefly explain the physical background of efficiency and fuel consumption of motor vehicles and present technical measures to improve it. Then it will explain how vehicle powertrain electrification can improve the efficiency of passenger cars and give an overview on the most important electrified powertrains concepts. After a brief description of their functioning and their key components a detailed analysis of their technical and economic perspectives will be performed.

5.1 Reducing Fuel consumption of motor vehicles

Fuel consumption has become an increasingly important parameter for passenger cars in the last years. Today reduction of fuel consumption is one of the most important fields of automotive R&D. In general reduction of fuel consumption can be approached in two ways: by reducing energy demand and losses in the vehicle and by improving the efficiency of the powertrain. A detailed view on the physical background of fuel consumption of passenger cars is given in Appendix A.

There are several ways propulsion energy consumption and energy losses can be reduced in a passenger car. The following sections will explain the main measures that can be taken:

5.1.1 Reduction of rolling resistance:

Rolling resistance can be reduced by using special tyres with extra low rolling resistance (Seiffert 2007a). Also the reduction of the vehicle mass helps to reduce rolling resistance.

5.1.2 Reduction of Aerodynamic drag:

Reduction of aerodynamic drag is mainly approached by reducing the aerodynamic drag coefficient c_w through improvements in the vehicle design. This has been an important subject in automotive R&D for the last decades and strong improvements could be achieved (Seiffert 2007b). Since the design of a passenger car is always a compromise between reduction of aerodynamic drag, usability of the car and crash safety the future improvement potential of the aerodynamics is limited and great further steps cannot be expected.

5.1.3 Reduction of vehicle mass:

Vehicle mass has strong effect on fuel consumption since it affects the rolling resistance, climbing resistance and acceleration resistance of the cars. In the last years high afford was made to reduce vehicle mass through intensified use of light-weight materials. Even though this has led to a reduction in the mass of some components (e.g. the chassis) (Timm & König 2008), the average mass of vehicles could not be reduced significantly. International analyses have shown that average curb weight of passenger cars has increased considerably in the last 2 to 3 decades (Berger et al. 2009) (MIT 2008). The main reason for this development is the fact that every new vehicle generation offered more safety and comfort features than the previous one making the vehicle heavier. This has led to negative feedback since heavier cars also require stronger and more powerful engines to maintain the same vehicle dynamic than the lighter predecessor (see chapter 2.4.2). This is why light-weight materials could only damp the mass increase but have not led to real reductions. In the future efficiency requirements could lead to an enhanced application of light-weight materials and to an abandonment of dispensable mass drivers which could cause an effective mass reduction of future car generations.

5.1.4 Improvement of engine efficiency

The low efficiency of passenger cars ($\approx$20 %) (see chapter 5.3) is mainly caused by the low efficiency of the internal combustion engine (ICE) which is still the dominating propulsion technology for passenger cars. Since engine efficiency directly affects fuel consumption (see equation (A-1) and (A-8)) in Appendix A, improvement of the engine efficiency has been in the focus of automotive R&D for many decades and considerable progress has been made (Christidis 2003). Today the technology reached a status where further improvements become increasingly difficult. However, experts believe that the potential of efficiency improvement in the motor is not yet exploited and further progress can be expected in the future (Nemry et al. 2008) Today, even small improvements in efficiency can only be achieved by raising the complexity of the engines. In the next decades stricter emission standards will require enforced exhaust gas after-treatment measures that will offset part of the efficiency gains in the engine (Wallentowitz et al. 2010). Another part of the efficiency gains could also be offset by the tendency that engine power is increased to improve vehicle dynamics in order to make the car more attractive to consumers.

The most recent approaches to improve engine efficiency are:

- direct injection and turbo charging of Diesel and Gasoline engines
- variable valve timing and friction reduction
- variable compression
- higher compression ratios and downsize of the motor
- decoupling of auxiliaries and running them electrically: Hereby the alternator, the oil and water pumps and even the camshafts can be decoupled from the crankshaft and can be controlled electrically. Thus they can run exactly according to the actual demand of the ICE which improves the overall efficiency of the engine.
- Cylinder cut-off
- Recuperation of waste heat
- New combustion processes

(cf. (Wallentowitz et al. 2010) (Nemry et al. 2008))

5.1.5 Drivetrain improvements

Theoretically, piston engines of motor vehicles can reach much higher efficiencies under ideal conditions. At optimal operation conditions the effective efficiency of the engine can reach up to 36 % in the case of gasoline engines and up to 43 % for diesel engines (Pischinger 2007). The problem is that the conditions an ICE has to face in a motor vehicle are far away from being ideal. The peak efficiency of the machine is reached in the optimal operating point, which means at one particular combination of engine speed and pressure. Once the machine deviates from this point the efficiency is decreasing drastically. In the case of a typical drive cycle of a passenger car the engine is deviating from this point very often resulting in this low efficiency. Advanced transmission systems enable the motor to run closer to the optimal operating point by flexibly shifting the transmission ratio (e.g. continuous variable transmission).

A recent and promising approach to improve the drivetrain is by electrification. A detailed view on this development will be given in chapter 5.2.

5.1.6 Electric Auxiliaries

The energy demand of auxiliaries also affects the fuel demand of the car (see equation (A-1 in Appendix A). The optimisation of their operation can also contribute to an improvement of vehicle efficiency. Usually, auxiliaries such as oil pump, power steering assistant, air condition etc. are linked to the drivetrain mechanically. This means that they always run at 100 % of their capacity even if less would be sufficient. When they are electrified, auxiliaries can run according to the requirements of the driving situation which saves energy.

5.2 Electrification of the vehicle powertrain

Electrification or hybridisation of the powertrain means mounting electric machines in the drivetrain to support the ICE and recuperate braking energy. Thereby, vehicle efficiency gets improved by two effects:

Firstly, the ICE is supported by the electric machines EM to run more efficiently. The low efficiency (ca. 20%) of cars with ICE is mainly founded in the fact that in normal driving cycles the engine is usually not running in its optimal operating point (for example during acceleration). In a hybrid powertrain the support of the electric motor enables the ICE to remain in the optimal operating point in most of the driving situations which increases its operation efficiency (cf. equation (A-1) and (A-8) in Appendix A).

Secondly, the electric machines can be used to recuperate breaking energy. In a conventional powertrain system during breaking the kinetic energy is converted into heat, which means that it is completely lost for the vehicles (exergy loss). In a hybrid system part of this energy can be recovered through the electric machines and stored in the battery. Later the energy can be used for the next acceleration phase. It is evident that the efficiency gain in comparison to conventional technologies increases with the number of acceleration and breaking phases in the reference cycle. That is why hybrid powertrains show there advantages especially in urban driving cycles. Due to the general trends of urbanisation and increasing traffic density the urban cycles become more and more important which will increase the relevance of hybridisation in the future.

Today, there are several concepts of hybridisation for passenger cars with different degrees of electrification. One parameter that is used to distinguish between hybrid systems is the ratio of the power of the electric machines (EM) compared to the power of the ICE. The stronger the power of the EM, the better it can support the ICE and the more braking energy can be recuperated which all makes the vehicle more efficient. Starting with relatively small EM, in so called *Mild Hybrid* configurations that can lead up to powertrain systems where the mechanical propulsion of the vehicle is provided exclusively by the EM and the ICE only drives an alternator to generate the electricity *(Series Hybrids)*. Therefore, hybrid technology is often seen as a first step on a pathway toward pure electric propulsion technologies.

In the following chapters the most important powertrain options will be explained briefly:

5.2.1 Conventional Drive (CD):

The term conventional drive (CD) is used for powertrain systems that are solely based on internal combustion engines without additional electric traction motors. They can be based on gasoline, diesel or CNG motors. The vast majority of vehicles today use CD propulsion systems.

5.2.2 Micro Hybrid:

These systems are closely related to CD systems. The powertrain is only slightly modified in comparison to the CD. The conventional electric starter and the alternator are replaced by one combined starter-alternator that is either mounted directly on the drivetrain or linked to it through a belt drive. Due to the higher number of starting processes higher battery capacity is required (Wallentowitz et al. 2010). With a micro hybrid system the ICE is switched off automatically when the car stops (for example at traffic lights). As soon as the driver releases the brake pedal the engine starts immediately. The electric machine is also used to recover breaking energy which is the major difference to ordinary start-stop systems. Due to the relatively low power of the generator only a small part of the energy can actually be recovered. Recovering braking energy also requires special energy management systems to control the energy flows. The system usually operates at low voltage levels (12-15V) that are also used for the on board grid. In some systems a second voltage level (42V) is used for higher recuperation. In city cycles a micro hybrid configuration can reduce fuel consumption by about 3-10% (Hofmann 2008). Since there is no electric propulsion involved the Micro Hybrid is sometimes considered not to be a real hybrid system but only a measure to raise efficiency of conventional drive systems. However, these systems have become popular in the last years and more and more car manufacturers offer them for their models.

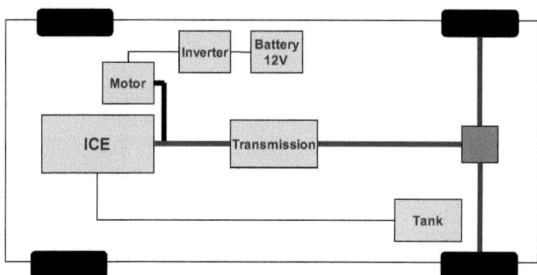

Figure 5-1: Micro Hybrid Powertrain System

5.2.3 Mild Hybrid

In Mild Hybrid System an electric motor (EM) is mounted on the crankshaft between the ICE and the transmission (see Figure 5-2). Since both the ICE and the EM are driving the wheels in parallel this configuration is also called *parallel hybrid*. The EM provides propulsion energy and supports the ICE in situations of high power demand, for example in acceleration phases. The EM has relatively low power compared to the ICE but high torque. That is why the main field of application of the EM is the acceleration phase where it provides the required peak power and torque which allows the ICE to remain in an efficient operation point. Pure electric operation is not foreseen in this concept (→ mild hybrid). Apart from acceleration support the system can also recover breaking energy and serves as a start stop system. In Mild Hybrids higher voltage levels (42-150V) are used to allow higher operation power and energy recuperation rates.

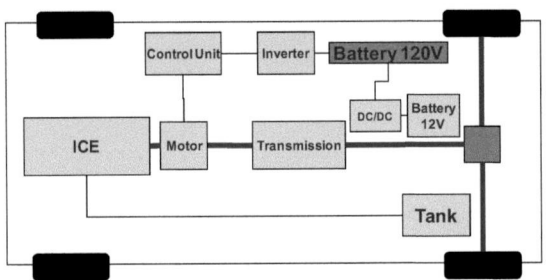

Figure 5-2: Mild Hybrid (parallel) Powertrain System

5.2.4 Full Hybrid:

In a full hybrid powertrain the ICE can be completely decoupled from the drive train by an extra clutch. The EM has much more power than in a mild hybrid which makes the electric mode a fully fletched driving mode. The powertrain architecture of a full hybrid can be similar to the one of the mild hybrid described above *(parallel hybrid)*. The only difference in this case is an extra clutch between EM and ICE permitting pure electric operation. However, the battery in a full hybrid is not designed for long distance electric driving and its standard operation mode is the combined operation with the electric machine supporting the ICE at acceleration and recuperating breaking energy.

In full hybrid systems higher voltage levels are used (more than 200 V) than in mild hybrid systems making the system able to recuperate breaking energy far more efficiently (Biermann 2008). The high torque of the electric machine allows strong acceleration support at all driving speeds. This operation requires a high voltage battery with high power flow. Today either NiMH or Li-Ion Batteries are used (see chapter 5.4.2).

Another powertrain architecture that is used for full hybrids is the so called *power split drive* (used for example in the Toyota Prius). A schematic view on the design of a power split hybrid propulsion system is given in Figure 5-3. In this systems there are two EM one serving as motor one as generator. The main advantage of this configuration is the fact that the battery can be recharged while driving. In this mode some of the kinetic energy of the ICE is bypassed to run the generator while the rest is used to run the vehicle. This makes the system very flexible and allows optimal operation of the ICE (Wandt 2008).

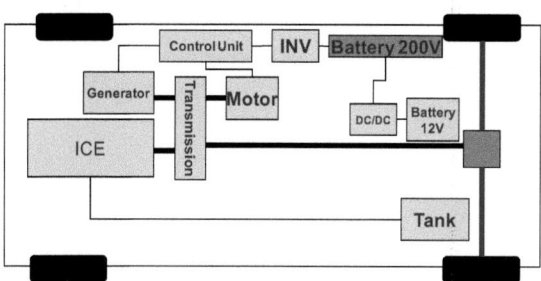

Figure 5-3: Full Hybrid Powertrain System (Power Split)

In this context it is important to explain another categorisation of hybrid systems: **parallel** and **series hybrid** systems.

In parallel systems both the ICE and the EM are mechanically linked to the drivetrain. They are operating parallel. In series hybrid systems the ICE is not mechanically linked to the drivetrain. It just runs a generator that produces electricity for the EM. A series hybrid configuration is depicted in Figure 5-5. The power split system (Figure 5-3) represents a combination of both systems that can run in both parallel and series driving mode.

5.2.5 Plug-In Hybrid (PHEV) – parallel & power split drive:

The main idea of plug-in-hybrid propulsion systems is to combine the advantages of electric vehicles (EV) and conventional drive (CD) vehicles in one system: High efficient and zero emission mobility of a pure electric car and the long driving ranges of conventional cars.

The powertrain of a PHEV can be quite similar to a full hybrid. The main difference is the capacity of the battery. PHEVs have higher battery capacities that can be recharged on the electric grid allowing pure electric operation. PHEV usually have lower electric driving ranges than pure electric cars. The electric range is just sufficient for every day trips (between 20 and 60 km). On longer trips the vehicle can switch to the ICE.

For PHEVs different powertrain configurations can be applied:

- *Series hybrid drive* architectures with the ICE permanently decoupled from the drive train. Such a system will be described in the following chapter 5.2.6.

- *Power Split drive* systems similar to the one of a full hybrid system. The main difference is the higher battery capacity that allows longer driving distances in pure electric mode.
- *Parallel hybrid drive* systems with an extra clutch between the ICE and the electric machine can also be used as PHEV (see Figure 5-4).

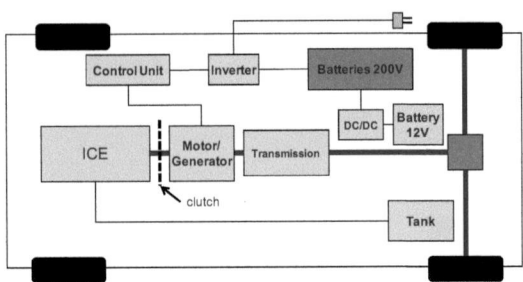

Figure 5-4: Plug-In-Hybrid

PHEVs can be operated in three main driving modes:

Charge sustaining mode: In this mode the vehicle is operated more or less like a full hybrid with the ICE and the EM in parallel while the state of charge of the battery is maintained.

Charge depleting mode: hereby, the vehicle runs in pure electric mode which means that the charge level of the battery is depleted until it falls below a certain level and the ICE is started.

Blended Mode: this mode is a combination of the two other modes. Depending on the driving situation the optimal mode is selected to maximize overall range. In the blended mode the battery is also depleted. Once it falls below a certain level the car has to switch into charge sustaining mode.

To tap the full potential of the blended mode the route should be predefined before starting so that the optimal strategy can be calculated. This requires a navigation system combined with intelligent software and if possible detailed information about the selected route (topology, speed limits etc.).

5.2.6 Plug-In Hybrid (PHEV) – Series Drive:

In a series hybrid electric vehicle (SHEV) the ICE is not linked to the drive train. It only drives an electric generator to produce electricity for the electric propulsion motors. The ICE and the electric machines are connected in series → series hybrid (see Figure 5-5). The system can also be seen as pure electric vehicle with an ICE as range extender (REX). Just as the PHEV the system combines the advantages of electric driving with the long range of ICE-based propulsion systems.

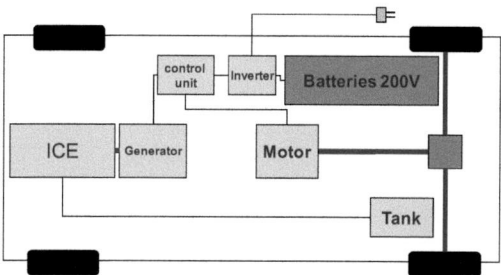

Figure 5-5: Series Hybrid System

The main driving mode of a series hybrid is the electric mode. The electric range should be high enough that most journeys can be realised in pure electric mode. For the typical user profile of Austrian passenger cars this means that the range should be between 40 and 80 km (see chapter 3.5). The electric range is also an economic question as the batteries are still the most costly component in these vehicles (see chapter 5.4.4).

There are two approaches to set the function and the dimension of the range extender ICE. One approach says that the vehicle should maintain its full driving capacities even in the series mode. This means that the ICE has to be strong enough to provide power close to the maximum power output of the electric propulsion system. In this case even long distance travelling at high speeds is possible which means that there are no disadvantages to the customer. This concept will be used in the Chevrolet Volt SHEV car which is planned to be introduced in the US in 2010 and in 2011 in Europe (as Opel Ampera) (GM 2010). This concept will be named *plug-in hybrid with series drive* (PHEV series drive) in this thesis.

The other approach only uses small ICE with output powers far below the maximum power of the electric propulsion system. In this case the ICE is mainly considered as a kind of emergency power supply for the rare case that the vehicle runs out of electric energy. Once running on the ICE the maximum driving capacities cannot be reached for longer time, which means for example that maximum driving speed is reduced. This concept will be called *battery electric vehicle with range extender* (BEV+REX) in this thesis.

5.2.7 Battery Electric Vehicle (BEV):

The battery electric vehicle (BEV) has only an electric drive train that receives all its electric energy from the batteries that have to be charged at the electric grid. That is why their overall range is relatively small compared to conventional cars. The lack of range and the high costs of batteries are seen as the major barriers to large scale market introduction today. To be acceptable for early adopters a BEV should have at least 150 – 200 km driving range which requires 20 to 40 kWh useable storage capacities (Aral 2009). Even with Lithium Ion Batteries the battery weight would still amount for 200 to 400 kg (see chapter 5.4).

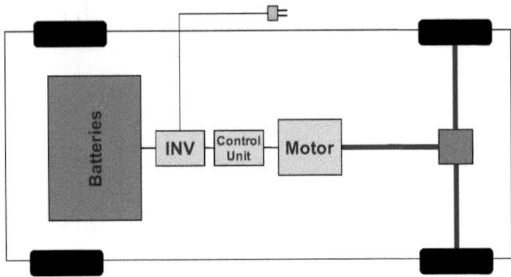

Figure 5-6: Battery Electric Vehicle

5.2.8 Fuel Cell Electric Vehicle (FCV):

Like the SHEV and BEV the FCV has a pure electric drive train. The main difference is the way the energy is stored. In the case of the FCV the energy is stored as hydrogen that is stored in the vehicle either compressed at high pressures (up to 700 bar) or as liquid at extremely low temperatures (<-253°C). In the fuel cell the electricity is generated from the hydrogen by a controlled reaction with oxygen. The main components in a fuel cell propulsion system are depicted in figure Figure 5-7. Apart from the fuel cell system a FCV usually requires a battery that acts as electric buffer storage to cover demand and supply peaks that are caused by acceleration and breaking phases during the driving cycle.

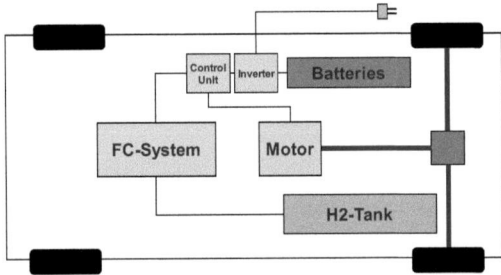

Figure 5-7: Fuel Cell Electric Vehicle

Fuel cell systems can be used as primary energy source or as a range extending system. When used as primary sources the fuel cell power output has to be as high as the nominal power of the electric drive system. In this case there would be only a small puffer battery comparable to the ones used in HEVs. The idea of this system is to run the vehicle exclusively on hydrogen just as HEV run on gasoline or Diesel. This represents the classical vision of fuel cell vehicles (FCV).

Another way to apply a fuel cell system in a vehicle is as range extender just like the ICE in a PHEV. In this work such powertrain systems are named Fuel Cell Plug-in Hybrids (FC-PHEV). The main driving mode of FC-PHEVs is on electricity from the batteries that are charged in the electric grid. Like in the SHEV the battery based driving range should be enough to drive most trips in this mode. The fuel cell system is only used when higher driving ranges are required.

5.3 Efficiency of automotive propulsion systems

When talking about energy efficiency of automotive propulsion systems it is important to define which part of the energy conversion chain is addressed. The view on the entire conversion chain from the production of the fuel until the conversion in mechanical energy in the propulsion system is called the well-to-wheel (WTW) view (see Figure 5-8). In particular when different types of fuels are involved it is necessary to consider the WTW view. The WTW chain is usually split up in two phases: the well-to-tank (WTT) part which covers the production of the fuel until it reaches the car and the tank-to-wheel (TTW) part that is covering energy conversion steps that are happening on board of the car. The TTW view is especially relevant when it comes to compare powertrain systems using the same fuels. When different fuels are involved the TTW view might be misleading and therefore should not be used in this case. This chapter will solely focus on the TTW balance. The energetic WTW performance of all analysed powertrain systems will be analysed in chapter 8.2.1.

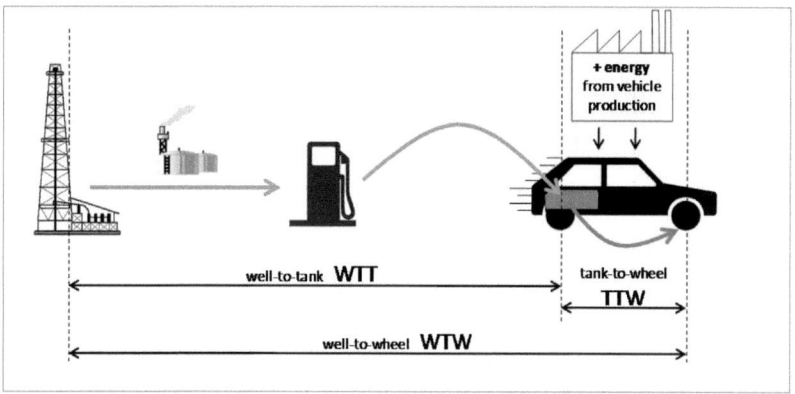

Figure 5-8: Definition of well-to-tank (WTT), tank-to-wheel (TTW) and well-to-wheel (WTW)

Figure 5-9 illustrates the TTW efficiency of a passenger car with conventional powertrain system. The diagram shows where the main exergy losses occur in the conversion of the chemical energy of the fuel to kinetic energy that drives the wheels of the car. Most losses occur in the engine, in the form of waste heat, friction, energy supply for auxiliaries and stand-by idle (Wurster et al. 2002).

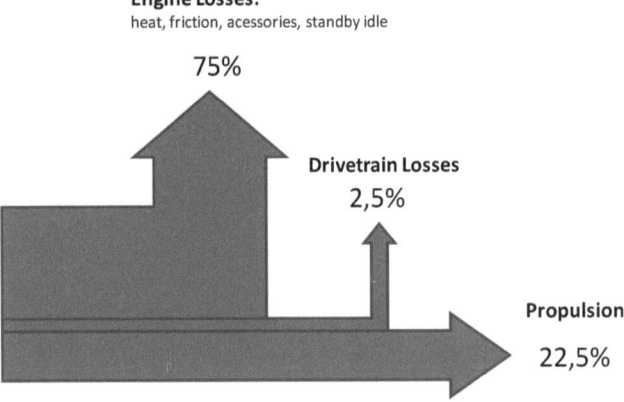

Figure 5-9: Tank-to-Wheel exergy losses in the a conventional passenger car propulsion system (TTW) Data Source: (Wurster et al. 2002)

As mentioned above significantly higher efficiencies could be reached with piston engines if they would run in their optimal operation point and if breaking energy would be recovered.

One objective of vehicle powertrain electrification is to improve the engine efficiency by holding it in the optimal operation point.

With their high torque that is already available at low speeds electric machines are very capable to support the engine in acceleration phases and their efficiency remains relatively constant in all operation points (Wallentowitz et al. 2010). The stronger the electric machine is dimensioned the better it can support the combustion engine to run efficiently and the more energy can be recovered. However, the coherence of electrification and energy efficiency is not linear and there is a maximum efficiency which is mainly given by the efficiency limit of the ICE (see figure Figure 5-10).

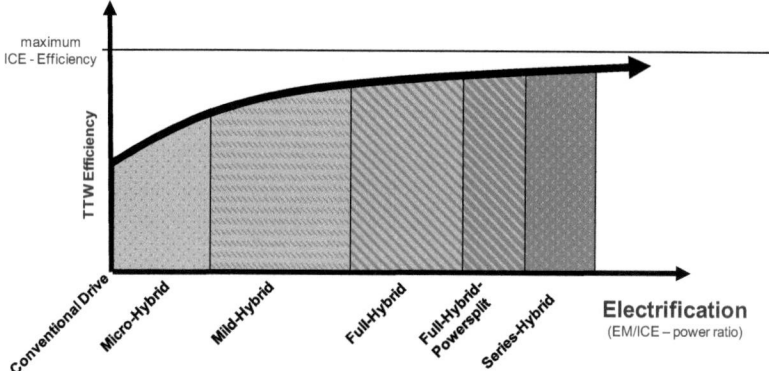

Figure 5-10: Efficiency improvement through electrification

In a hybrid powertrain the efficiency of the internal combustion engines can be increased from around 25 % to over to 30 % (Christidis et al. 2005). The possibility of energy recuperation of HEV systems further improves the vehicle efficiency enabling a TTW efficiency of up to 28 % (see Figure 5-11).

In the series hybrid (SHEV) configuration the ICE runs in its optimal operating point with the corresponding high efficiency (36 % for gasoline engines (Pischinger 2007)) (see chapter 5.2.6). However, without a fixed mechanical drive there are conversion losses in the chain: mechanical energy → electricity → mechanical energy. Depending on the driving cycle, part of the energy flow does not directly run through the generator to the motor but takes the way through the battery causing further losses. The high efficiency of modern battery technologies and modern electric machines (usually Permanently Magnetised Synchronous Machines - PMSM) make hybrid concepts more and more attractive since losses in the electric drivetrain are minimized.

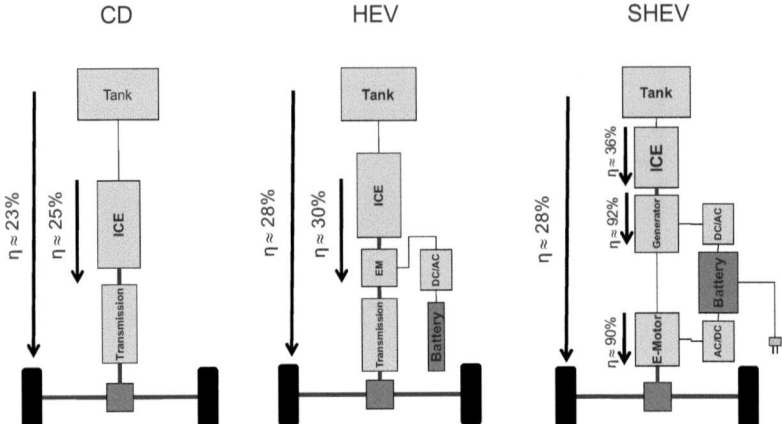

Figure 5-11: TTW Efficiency of ICE-based powertrain systems

The improved efficiency of electric drive components is also pushing the performance of battery electric vehicles. Since energy storage capacity in these cars is limited the efficient use is crucial. Battery electric cars reach TTW efficiencies of around 75 %. The main losses in electric powertrain system, apart from those in the motor ($\eta \approx 90$ % (Wallentowitz et al. 2010)), are the discharge losses from the battery ($\eta \approx 90$ % for Li-Ion (Matheys & Autenboer 2005)) and losses generated at DC/AC conversion ($\eta \approx 97$ % (Campanari et al. 2009)).

The fuel cell vehicle (FCV) also benefits from the improvements in electric drive components. However, their efficiency is strongly affected by the efficiency of the main energy conversion step, the transformation of hydrogen into electricity in the fuel cell. Today there are fuel cell systems with efficiencies of up to 65 % (Kojima & Morita 2010) and even 70 % are considered feasible (Kloess et al. 2009).

For a detailed overview on losses in different powertrain systems see table Table A-3 in the appendix.

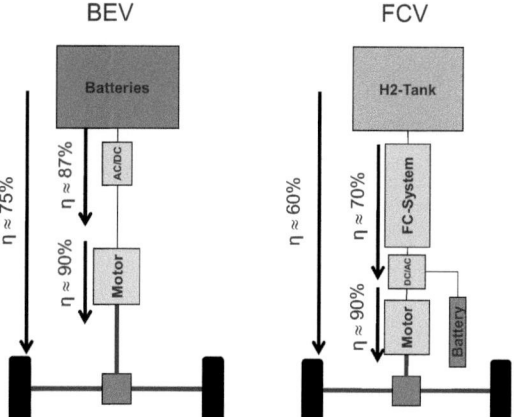

Figure 5-12: TTW Efficiency of Battery Electric Vehicle (BEV) and Fuel Cell Vehicle (FCV)

From a TTW perspective BEVs and FCVs have much higher efficiencies than any ICE based car even if they are hybridised. However, TTW comparison with ICEs is not admissible since different fuels are used. For a correct assessment the whole energy conversion chain (from Well to Wheel WTW) has to be considered. A comparison of the well-to-wheel performance of all powertrain systems considering different fuel options will be given in chapter 8.2.1.

5.3.1 Vehicle Efficiencies in the analysis

For a correct determination of fuel consumption in test cycles simulation tools are required.

The efficiencies of the different propulsion systems used in this analysis were determined by Researchers of AVL List, a company specialised on automotive Research and Development. The results are tank-to-wheel (TTW) efficiencies of all vehicles and powertrain systems and their corresponding fuel consumptions and greenhouse gas emissions for a combined test cycle (ARTEMIS and NEDC). The efficiencies of the technical status 2010 were determined with vehicle simulation tools developed by AVL. Experts from AVL also estimated the improvement potential of each technology to determine the 2050 status of the technologies (a closer description can be found in the project report: see (Kloess et al. 2009)).

Figure 5-13 indicates that vehicle efficiency is increasing with the degree of electrification. The most efficient vehicle is the BEV with a TTW-efficiency of over 70 % (2010 technology status). Concerning the PHEV and the SHEV it has to be considered that the depicted TTW efficiency is in series mode which means that the system is running on the ICE. In pure electric mode the efficiency would be as high as the BEV. The data for the 2050 technology status show that all systems have considerable potential for improvements with slightly higher improvement potential for electrified systems than for conventional systems that are mainly based on the mature technology of the ICE.

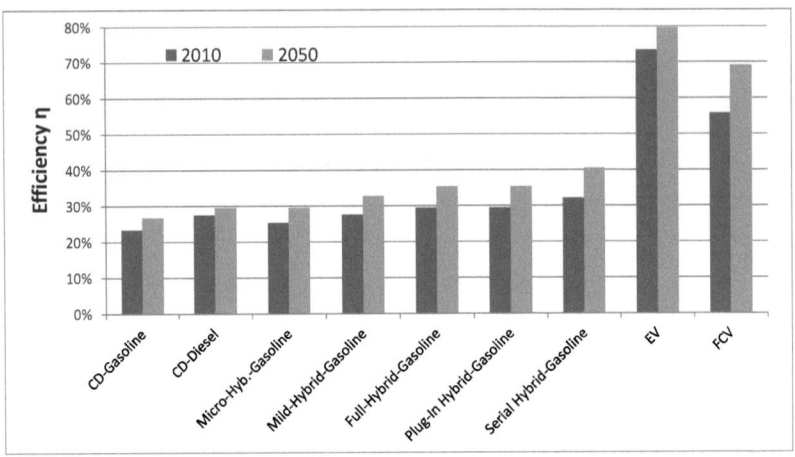

Figure 5-13: TTW Vehicles Efficiency 2010 and 2050 (Data Source: ELEK-TRA Project (Kloess et al. 2009))

5.4 Electricity storage systems for Electric Cars

Electric propulsion systems are a capable alternative to ICE based options in many fields. They have no direct emissions, they are highly efficient, they have low noise and they offer an excellent driving performance in terms of acceleration and driving dynamics.

The main handicap of battery electric vehicles is the limited driving range and the low charging speed. Today, the expectations of the customer concerning these two parameters are strongly affected by the specifications offered by ICE based vehicles today (see chapter 4.6.1). Every alternative propulsion technology has to meet these expectations, to compete.

That is why the electricity storage has turned out to be one key problem for electrified cars. Storing enough electric energy in the car to provide acceptable driving ranges is a problem that has not been solved satisfactory so far. There are different approaches to face this problem. The best known technologies are electrochemical batteries, but there are also other electricity storage options like hydrogen based fuel cells or capacitors. Fuel cells have been subject to intense R&D efforts of automobile companies and suppliers in the past two decades. Nevertheless, the technology didn't reach a status where commercialisation seemed feasible because of technical and economic reasons (see chapter

4.5.3). The failure of the fuel cell turned automotive R&D focus back on electrochemical electricity storage technologies.

Electrochemical batteries are the standard electric storage systems for many mobile applications. They are commonly used in mobile phones, computers and any kind of mobile consumer electronics. For these devices they meet customer expectations quite well and their performance characteristics have improved continuously.

However, the driving ranges that consumers are used to have today cannot even be reached with most recent battery technologies. For an electric range comparable to a gasoline or diesel car (700-1000 km with one charge) the battery weight would exceed 1 t even if the most advanced battery technology was used (see chapter 5.4.7).

With today's technology status it is impossible to achieve comparable driving ranges like conventional cars with a BEV. Reaching these driving ranges either requires a revolution in battery technology or other electricity storage concepts. Most of these alternative concepts are based on the idea of storing the energy in another medium and generate the electricity for the drivetrain on board.

Thereby, the high energy density of these energy carriers can be utilized for storage while the vehicle is still running fully electric. For example they can use internal combustion engines running on liquid or gaseous hydrocarbons to drive an electric generator. These so called range extenders (REX) are also applied in series hybrid powertrain systems (see chapter 5.2.6). Another option could be hydrogen based fuel cell systems. In chapter 5.4.5 and 5.4.6 the functioning of range extenders will be described in more detail.

Even with range extenders these cars are primarily designed to run on battery for most of the time, which points out the fact that batteries are the key components for the electrification of passenger car transport. In the following chapters an overview on the most important technical and economic parameters and factors will be given including both technical and economical parameters.

5.4.1 Relevant Parameters for the assessment of batteries

In Automotive applications batteries have to meet different requirements that require different characteristics. For the comparison of storage technologies some key parameters are used:

5.4.1.1   Gravimetric and volumetric energy density:

The corresponding unit for this parameter is Wh kg^{-1} respectively Wh l^{-1}. The parameters define the energy that can be stored in a battery system in relation to its weight or volume. The specific energy density is important when long time constant discharge is required as it is the case for BEVs. Both are key parameters for the electric range of a BEV or PHEV.

5.4.1.2 Gravimetric and volumetric power density:

For applications that require high power in a short time the gravimetric and volumetric power density are important. The unit for this parameter are W kg^{-1} respectively W l^{-1}. HEV are typical field of application where these characteristics are required. In these propulsion systems the battery has to provide or to absorb high energy flows for a short time for example in acceleration phases or for recuperation of breaking energy.

5.4.1.3 Efficiency:

Efficiency is another key parameter for batteries. It determines the relation between the electric energy input (when charging the battery) and the electric energy output when discharging the battery.

$$\eta = \frac{E_{in}}{E_{out}} \quad [\%] \tag{5-1}$$

η ... efficiency
E_{in} ... energy input
E_{out} ... energy output

The losses that occur in the battery are mostly heat losses at inner resistances. The efficiency is one key criterion for both hybrid and electric vehicles. In both application fields the battery efficiency is crucial for the recuperation efficiency of the car. For traction batteries high discharge efficiency is important to optimally exploit the energy stored in the vehicle and thereby maximize vehicle range. That is why efficiencies of over 90 % are required for BEVs and HEVs (Köhler 2007).

5.4.1.4 Cycle stability/durability & Calendar life:

For the durability of batteries the cycle stability is an important parameter. The cycle durability is given in number of cycles until failure of the cell. It has to be differentiated between deep and shallow cycles, also named charge depleting and charge sustaining cycles. For hybrid batteries the number of charge sustaining cycle is more important while for BEV and PHEV the charge depleting cycles are relevant too.

Apart from number of cycles the status of the battery is also affected by the calendric life. The calendar life describes the degradation of the cell due to undesired chemical reactions in the cell because of limited thermal stability. Thereby, the degradation is accelerated by high ambient temperature.

One great challenge for the future will be to develop batteries for BEVs and HEVs with enough durability and calendar life to last the entire car life (10 years or 200 000 km) (Köhler 2007).

5.4.1.5 Nominal and usable capacity of the cells

The nominal capacity of the batteries defines the electric energy that can be stored in the battery. However, in practice not the entire nominal capacity of the cell is used in cycle operation but the battery is operated in a certain range that depends on the characteristics of the particular cell technology. To keep the state of charge (SOC) of the battery within this range battery management systems are used that control the minimal and maximal depth of discharge (DOD). This is mainly done because deep discharge and overcharge compromise battery life.

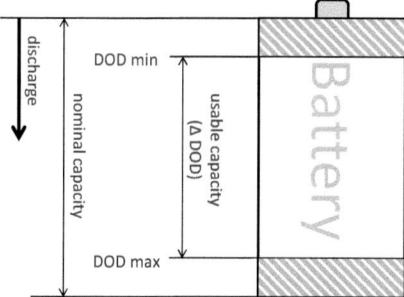

Figure 5-14: Nominal and usable capacity of batteries

5.4.2 Battery Technologies

There are roughly four groups of battery technologies that are used for automotive purposes:
- Lead-based Batteries
- Alkaline Batteries
- Lithium Based Batteries
- High Temperature Batteries

All these technologies have different characteristics with pros and cons. This chapter will give a brief overview on the most important technologies that are used for automotive applications and point out their strengths and weaknesses. A special focus will also be on the historic development of the technical key parameters to find out what progress has been made in the last three decades on the field of electrochemical electricity storage technologies.

5.4.2.1 Lead Acid Batteries (PbAc):

PbAc batteries are the most common battery type for automotive application today. Normally they are used as starter batteries and as distributor for the on board electricity grid. However, they have also been used as traction batteries in the past. When used for traction purposes a different design is applied than for starters as they have to be resistant to deep discharge cycles (Passier et al. 2007).

Lead Acid batteries are technologically mature since they have been in use in different applications for decades and their costs are low in comparison to other technologies (see chapter 5.4.4). The disadvantages of these batteries are their low energy and power density mainly caused by the fact that the major component in the battery is lead.

One famous car using lead acid traction batteries was the EV1 produced by General Motors in the 1990ies with a battery weight of approximately 530 kg holding 18.7 kWh of energy and permitting a driving range of 90 – 140 km (Helmers 2009).

5.4.2.2 Nickel Cadmium Batteries (NiCd)

These alkaline type batteries were used for BEVs in the 1990ies. They have considerably higher energy densities than lead batteries but they also were more costly. The production of these cells was faced out when superior technologies (e.g. NiMh) emerged on the market and also for environmental reasons (Passier et al. 2007).

5.4.2.3 Nickel Metal Hydride Batteries (NiMh):

These batteries were developed to replace the NiCd batteries which were considered harmful because of their cadmium content. They are also superior to NiCd batteries in terms of their energy density and replaced them very soon for all kinds of mobile applications. They were also the first to be used in large scale in HEVs (e.g. Toyota Prius). Their main advantages are their high energy- and power density, their robustness and the reduced memory effect. Due to their high power density they are especially interesting for HEVs (see Table 5-1). Also their energy density is about two times higher than for lead acid batteries (Passier et al. 2007). One critical disadvantage is their low efficiency. That is why they are being replaced by lithium Ion batteries in portable applications and are also likely to be phased out for automotive application within the next years.

5.4.2.4 Lithium Ion Batteries (Li-Ion)

The most recent battery technology is the lithium Ion battery. They have been used in cell phones, computers and consumer electronics for more than a decade and now they are considered for automotive applications. Their characteristics are superior to most other types of batteries. Especially their high power density makes them an interesting option for HEVs. Due to their high energy density and their high efficiency they are also very promising for PHEVs and BEVs (see Table 5-1 and Figure 5-16).

There are different chemistries of Lithium Batteries with different characteristics and at different stages of development (Passier et al. 2007). Some very promising chemistries are still in the development stage which points up the remaining potential for future improvements in this technology (Kalhammer et al. 2007).

5.4.2.5 High temperature batteries: Sodium-Nickel Chlorid (NaNiCl2) ZEBRA

These batteries were developed as traction batteries for electric vehicles in the 1990ies. They use an operating temperature of around 250-300°C which has to be maintained permanently to operate the vehicle even when the vehicle is not used (see (Passier et al. 2007) and (Köhler 2007)). This is why the battery is not adequate for typical user patterns of private passenger cars, but more appropriate for fleet use. They have relatively high energy densities but modest power densities (see Table 5-1). Today they are not considered as a promising technology option for future PHEVs or BEVs.

Table 5-1: Characteristics and Cost of different Battery Technologies (data sources: (Passier et al. 2007)(Matheys & Autenboer 2005)(M. Conte et al. 2004)(van Vliet et al. 2010))

			Energy Density [Wh/kg]	[Wh/l]	Power density [W/kg]	efficiency [%]	temperature range [°C]	Cycles	Cost [€/kWh]	Cost [€/kW]	Source
Lead Based	Lead Acid	flooded	25-50	60-100	140-350	70-75	20-40	200-1500	100-190		Passier et al 2007. / IEA Hybrid Outlook
		VRLA	30-40	80-100	140-300	80-85	20-40	300-1000	100-190		Passier et al 2007. / IEA Hybrid Outlook
		compressed	40-50	100	140-250	70-85	20-40	800-1500	35-50		Passier et al 2007. / IEA Hybrid Outlook
	Power		40		250	80-85		500	116-151		Matheys et al. SUBAT Report 2005
			30	80	450			3000	150		Conte et al. 2004
	Energy		40		250			500		12-15	Matheys et al. SUBAT Report 2005
			37	120	200			500	120-150		Conte et al. 2004
Alkaline	NiCd										
	Power		25-40	130	500	70-75	-40 - 50	800-1500	400-1000		Passier et al. 2007. / IEA Hybrid Outlook
			30		500				490-720		Matheys et al. SUBAT Report 2005
	Energy		40-50	130	120-350	70-75	-40 - 50	800-1500	400-1000		Passier et al. 2007. / IEA Hybrid Outlook
			60		240	70-75	0-40	1350		52-54	Matheys et al. SUBAT Report 2005
			55	110	400			1500			Conte et al. 2004
	NiZn										
	Energy		60-80	200-300	500-1000	60-65	0-40	200-1000	500-800		Passier et al. 2007. / IEA Hybrid Outlook
			70-80	150	200			1200	300 (@100k units/a)		Conte et al. 2004
	NiMh										
	Power		40-55	80-200	500-1400	70-80		500-2000	400-1000		Passier et al. 2007. / IEA Hybrid Outlook
			55		1500					46-60	Matheys et al. SUBAT Report 2005
			35-40	70-80	200			1200	300 (@100k units/a)		Conte et al. 2004
	Energy		60-80	200-350	200-600	70-80		500-2000	400-1000		Passier et al. 2007. / IEA Hybrid Outlook
			70		350	70		1350			Matheys et al. SUBAT Report 2005
			70-80	150	650			100000	400-450 (@100k units/a)		Conte et al. 2004
Lithium Based	Lithium Ion										
	Power		70-130	150-450	600-3000	85-90		800-1500	700-2000		Passier et al. 2007. / IEA Hybrid Outlook
			70		2000					44-52	Matheys et al. SUBAT Report 2005
			90	85	1400			300k	500 (@100k units/a)		Conte et al. 2004
	Energy		110-220	150-450	200-600	85-90		800-1500	150-500		Passier et al. 2007. / IEA Hybrid Outlook
			125		400	90		1000	700-860		Matheys et al. SUBAT Report 2005
			125	210	370			750	300 (@100k units/a)		Conte et al. 2004
			90-110		1200-3000				1000-1600		Vliet et all 2010
	Lithium Polymer		100-180	100	300-500	90-95	-110	300-1000	300-500		Passier et al. 2007. / IEA Hybrid Outlook
	Na-NiCl										
	energy		94	140	169			1000	345 (@100k units/a)		Conte et al. 2004
			125	200		90-95		1000	450-500		Matheys et al., SUBAT Report 2005

5.4.3 Technological Progress of Batteries

Historically batteries have been the major weakness of electric cars. In order to permit sufficient electric range the specific energy is one of the key parameters for energy storage technologies. At the very beginning of the history of electric vehicles at the beginning of the twentieth century the applied Lead based batteries reached specific energy between 10 and 20 Wh kg^{-1}. In the following decades very little progress in battery technology was made and it took until the 1990ies to reach energy densities higher than 50 Wh kg^{-1} (see Figure 5-15). In the last two decades considerable progress in the field of energy density of batteries has been made and today Lithium-based batteries reach energy densities of up to 150 Wh kg^{-1}.

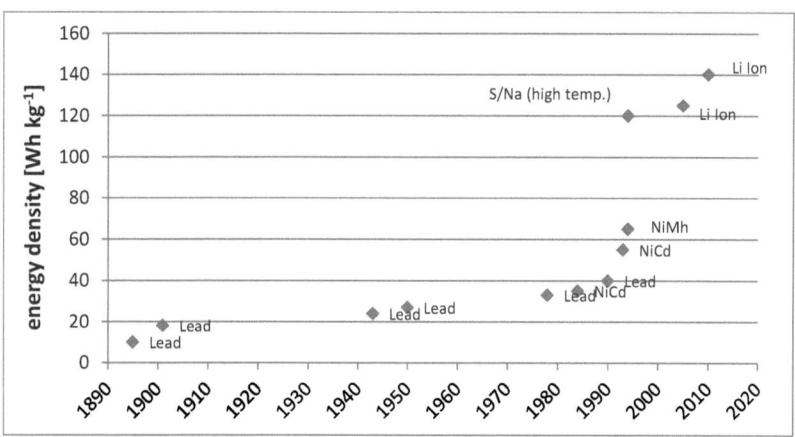

Figure 5-15 Development of specific energy of Batteries – Sources: (Cowan & Hultén 1996); (W. Fischer 1994); (Matheys & Autenboer 2005); (Burke et al. 2010)

Nevertheless, even the most recent technologies are not able to fulfil the requirements of an energy storage system for an electric car sufficiently. A battery for an EV has to have high energy density and efficiency to permit sufficient driving range and it has to offer good durability in terms of both calendar life and cycle life. Furthermore, thermal requirements should be compatible with conditions found in an automobile. The standards that these batteries have to fulfil to be adequate for large scale automotive application are summarized in the so called USABC-goals for battery development (USABC 2009)(USABC 2009). The United States Advanced Battery Consortium (USABC) is a sub-consortium of the United States Council of Automotive Research (USCAR) which was founded to support activities that lead to advanced automotive propulsion. Since its foundation the USABC kept setting mid-term and long-term development goals for automotive batteries. The present goals are illustrated in Table 5-2.

Table 5-2: USABC Battery Goals (source: (USABC 2009))

		Specific Energy [Wh/kg]	Energy Density [Wh/l]	Specific Power [W/kg]	efficiency [%]	Cycles	Calendar life [years]	Cost [$/kWh]
BEV	minimum goal	150	230	150-300	90	1000	10	150
	long-term goal	200	300	200-400	90	1000	10	100
PHEV	high power/energy ratio (PHEV-10)	100		930	90	300k	15	300
	high energy/power ratio (PHEV-40)	140		380	90	300k	15	200
HEV				625	90	300k	15	2000

So far there was no single technology that could meet all these requirements sufficiently. Even if some technologies had the capacities to fulfil one requirement they couldn't fulfil others.

In the last years great progress in lithium Ion battery technology has been made and the technology is considered a potential candidate to fulfil these goals. As illustrated in Figure 5-16 lithium Ion batteries have the potential to fulfil even the goals for BEVs. By now (2010) the full potential of these batteries could not be completely exploited due to unsolved technological problems (Passier et al. 2007) & (Kalhammer et al. 2007). Today, the minimum battery goals that were defined by the USABC for the application in PHEVs are already met (see Figure 5-16).

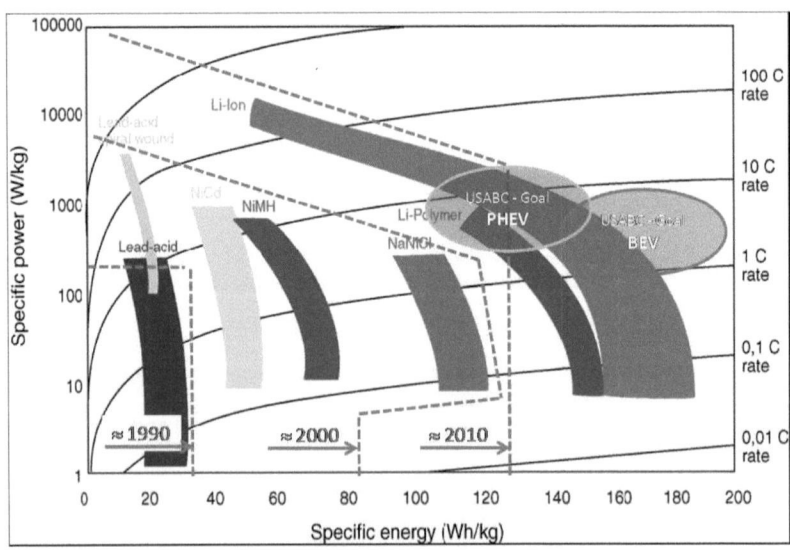

Figure 5-16: Technical Evolution of Batteries & USABC Development Goals (adabted from SUBAT (Matheys & Autenboer 2005))

The development of Li-Ion batteries is subject to intense R&D efforts. In the future technological progress could bring Li-Ion batteries closer to the long term goals. However, past experience has shown that short-term progress can easily be overestimated. For example at the beginning of the 1990ies, when USABC battery goals were set for the first time the long term goal for energy density (200 Wh kg^{-1}) was already the same as today's long term goals (W. Fischer 1994). Obviously the long term goals of the early 1990ies implied a technological breakthrough that never happened. The same is happening today where real technology specifications are still significantly below the long-term goals.

However, taking a look at the long-term development of battery systems (see Figure 5-15) it shows that progress in battery development has accelerated in the last two decades. Today the potential of Li-Ion batteries is largely exploited, but there is intense research on new promising technologies like LiFePO$_4$ (Howell 2009) or Lithium- and Sodium-Air (Peled et al. n.d.), which could offer further leaps in energy density in the next decades.

5.4.4 Specific costs of batteries

Another important parameter for electricity storage systems is their cost. Today, battery cost is one of the major barriers to overcome for all kinds of electrified vehicles. Batteries are a sever cost driver for hybrid cars and in particular for battery electric cars. Also in the past the high cost of batteries was a major barrier to the market introduction of electric cars (see (Sarnes 1992) , (U. Wagner 1988)).

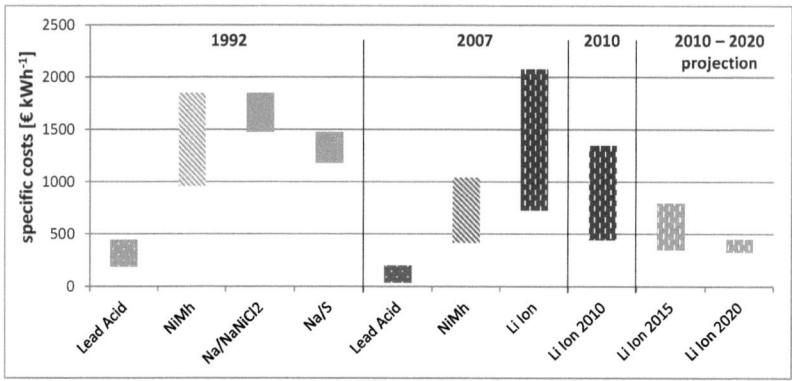

Figure 5-17: Real$_{2010}$ Specific costs of battery technologies for electric cars. (Sources: (Braess 1992); (Passier et al. 2007); (J. F. Miller 2010))

Figure 5-17 compares the cost of battery systems in the year 1992 with the current costs of relevant battery technologies. In 1992 NiMh batteries were the most recent technology specific costs of 1400-1800 € kWh^{-1}. High temperature batteries with high energy densities like Na/NaNiCl2 and Na/S had about the same price back then (cf. (Braess 1992)). Up to 2007 the cost of NiMh batteries decreased by a factor of 2-3. For Li-Ion batteries there is a high range of costs depending on the characteristics of the cells and the applied materials (cf. (Kalhammer et al. 2007) & (Passier et al. 2007)). Up to 2010 costs of Li-Ion batteries have reduced considerably and further reductions are projected for the upcoming decade (J. F. Miller 2010) (see Figure 5-17). Similar estimations are made by other experts: For example (Chalk & J. F. Miller 2006) assumes that 150 $ kWh^{-1} can be achieved even in a short term through production up-scaling and (MIT 2008) assumed 250 $ kWh^{-1} to be feasible (see chapter 5.4.4). Comparing the present cost status with the minimum requirements defined by the USABC (Table 5-2) it shows that there still remains a considerable gap in actual costs and the defined goals.

In chapter 5.6 a cost estimation of future battery cost 2010-2050 will be presented.

5.4.5 Range Extenders

The gap between electricity storage requirements and the actual state of technology calls for other ways of electricity storage. One of this approaches are range extenders (REX). The main functioning of REXs was described in chapter 5.2. In this chapter the topic will be approached from an energetic perspective.

The basic idea of range extenders for electric cars is to store energy in another form and transform it in electricity on board of the car. The most common approach is to store the energy as liquid hydrocarbons (e.g. gasoline or diesel) and use a combustion engine and a generator to transform it into electricity. Because of the excellent storability and the good availability of gasoline and diesel ICEs are the technology of choice for range extenders today. By using a range extender the driving range of the vehicle is not limited by the battery capacity and allows comparable overall driving ranges as ICE-based cars and permit fast refuelling. Another advantage of range extenders is the fact that the costly battery can be kept smaller. With a smaller battery, costs can be cut considerably while according to the typical user pattern most of the trips can still be done in electric mode (see chapter 2.5). The schematic illustration of battery charge depletion of an EV with REX is given in Figure 5-18. The first kilometres are always covered in the fully electric mode. Once the state of charge falls below a certain level the REX is activated and keeps the battery charge on a constant level. To permit 100 % system power for the entire vehicle driving range an energy reserve has to be considered. Therefore, the starting level of the REX has to be above the minimum charge level.

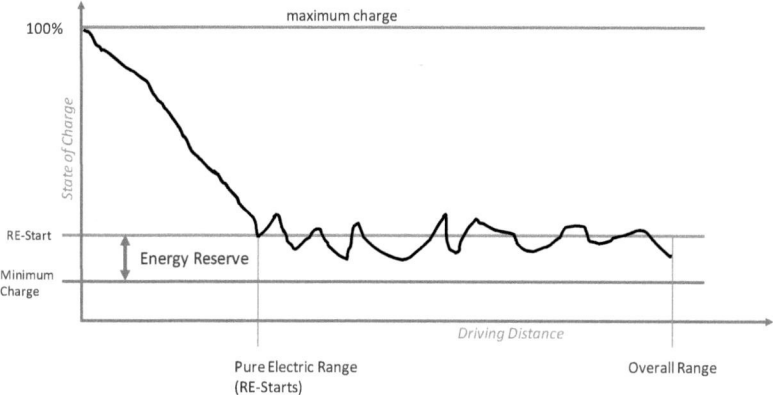

Figure 5-18: Battery depletion in an EV with range extender (Source: adapted from AVL (Sorger et al. 2009))

The parallel installation of electric drive components and an ICE makes the propulsion system more complex and costly, which is also the main disadvantage of the concept. However when used as a REX the ICE does not need to provide 100 % of the car's system power. It only provides more or less a base load supply that is high enough to permit long distance driving at motorway speeds (e.g. 120 km/h). Therefore, a smaller engine can be used than in conventional cars. In case of higher power requirements, for example in acceleration phases, additional power of the battery is used. At lower demand phases the battery can be recharged during driving, so that 100% system power will be available in all driving situations.

Hence smaller engines with lower displacement can be used. Figure 5-19 shows two types of range extenders for electric cars: One 2 cylinder four-stroke piston engine and one rotary engine concept (Sorger et al. 2009). Both have a very compact design and are integrated in one module together with the electric generator.

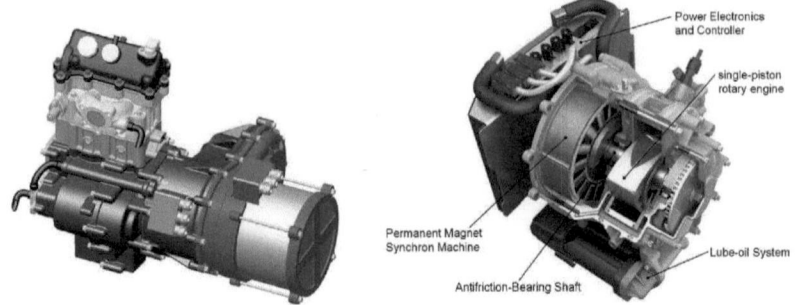

Figure 5-19: ICE-Range Extenders with piston- and rotary engines (Source: AVL (Sorger et al. 2009))

5.4.6 Fuel Cells

Fuel cells have been a candidate for an alternative propulsion technology for passenger cars for more than two decades (see chapter 4.5.3). The basic idea of using fuel cells in a car is to be able to store the energy in the vehicle in another form and converting it into electricity on board instead of using heavy batteries. The fuel cell usually runs on pure hydrogen that is stored on board either in liquid or in gaseous state. There were also attempts to use hydrocarbon fuels and on-board reformers to separate the hydrogen. However, it turned out that the efficiency gain would not justify the high complexity of these concepts (Wurster et al. 2002).

Today fuel cell related R&D is focussed on the use of pure hydrogen. Due to the high energy density of hydrogen and the high efficiency of the fuel cell (see chapter 5.3) high driving ranges can be reached that are comparable to conventional cars. Furthermore, the refuelling process takes just a few minutes which makes the entire concept more compatible to the way cars are fuelled today. However, hydrogen on-board storage remains a technical problem that is not solved satisfactory yet. The standard storage technologies for hydrogen are either in gaseous state at pressures between 350 and 700 bar, or in liquid state at cryogen temperatures (-253°C) (Helmolt & Eberle 2007). Both storage technologies are energy intensive and therefore create significant losses in the energy conversion chain (see chapter 5.3 and 8.2.1).

The basic idea of fuel cell cars was to use the fuel cell as the main propulsion system. These vehicles just have small batteries with high specific power used for energy recuperation just like in hybrid cars. In this case the power of the fuel cell systems has to correspond to the maximum system power.

Another field of application for fuel cells is to use it as a range extender for electric cars. Like in ICE-based range extenders, the energy is stored on board in another form, namely hydrogen, with higher energy densities (see chapter

5.4.7). In this case the fuel cell can be significantly smaller than in pure fuel cell cars which would reduce costs considerably. The main advantage over ICE based range extenders is the fact that the car is running with zero emissions even in the range extender mode.

5.4.7 Electric range and curb weight in electric propulsion systems

Providing sufficient driving range for pure electric propulsion systems is still an unsolved problem. As mentioned in the previous chapters batteries have still relatively low specific energy densities in comparison to hydrocarbon fuels. Figure 5-20 illustrates the difference in energy density between gasoline and a stat of the art lithium ion battery (100 Wh kg^{-1}) and shows that there is a factor of 118 between the two storage concepts. When the better efficiency of the electric drivetrain compared to the ICE is considered the factor is still 42 (see Figure 5-21). This points out why higher electric driving ranges also means a considerable increase of vehicle mass.

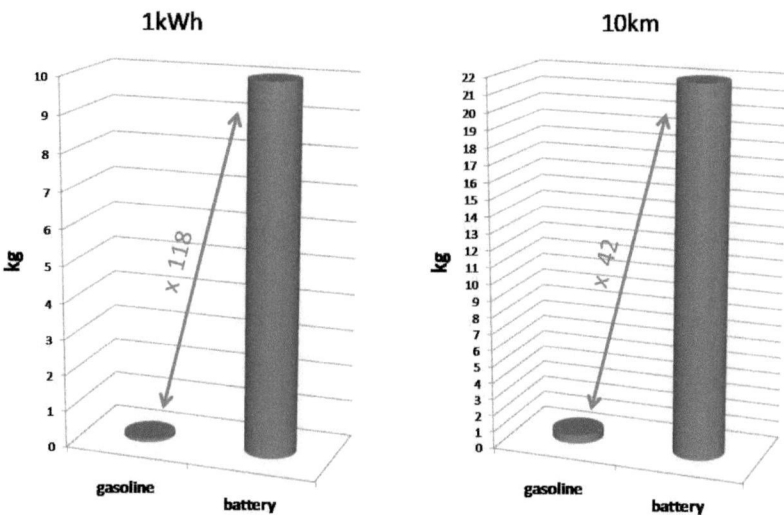

Figure 5-20: Required mass to store 1 kWh on board

Figure 5-21: Required mass to store energy for 10 km

Total mass has a critical impact on the fuel consumption of a car (see Appendix A). In an electric car higher range means more batteries and consequently higher mass. Figure 5-22 illustrates how the mass of a battery electric car increases with electric driving range. The specific energy density of the battery system is assumed to be 100 Wh kg^{-1} which is technically feasible with Li-Ion technology today (see chapter 5.4.2). The mass increase is non-linear as higher vehicle mass also leads to higher fuel consumption. In this case only the additional battery mass is considered. In practice the mass would be further increased by necessary reinforcements of the vehicle structure and possibly a stronger propulsion system. Consequently, there are technical limits in electric range of BEVs today and comparable ranges of conventional cars cannot be reached with current battery technology.

A comparison with the weight structure of conventional drive (CD) cars using gasoline or diesel based ICEs underlines the main deficit of BEVs. Due to the higher energy density and good storability of liquid hydrocarbon fuels the mass increase caused by higher ranges is practically negligible in CD cars (see Figure 5-23) (cf. (Kloess 2010)).

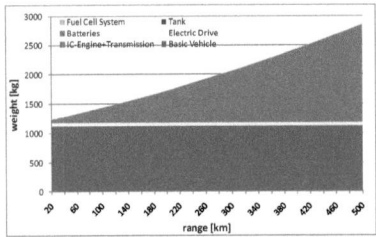
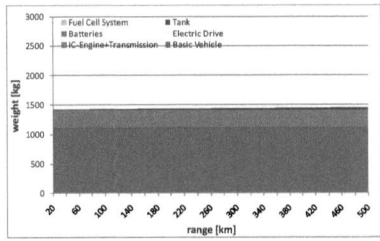

Figure 5-22: Impact of increasing range on vehicle mass (BEV)

Figure 5-23: Impact of increasing range on vehicle mass (CD)

The mass comparison of the two powertrain technologies shows that high ranges, like those consumers are used to have today, are not feasible with current electrochemical storage systems. Even if the specific energy density could be improved to around 150 Wh kg^{-1} electric ranges over 400 km would still be unfeasible.

As mentioned above a possible solution for electric powertrain systems to reach higher ranges are range extenders. With ICE based range extenders the range is only limited by the size of the fuel tank which means that they would become equal to CD cars.

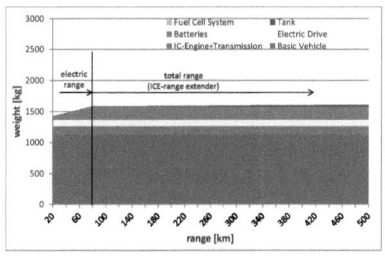
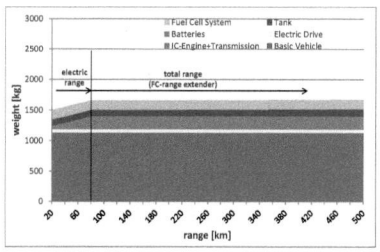

Figure 5-24: Impact of increasing range on vehicle mass (PHEV-80km series drive)

Figure 5-25: Impact of increasing range on vehicle mass (FC-PHEV-80km)

Figure 5-24 and Figure 5-25 illustrate the impact of driving range on vehicle mass when range extenders are used (ICE based and hydrogen fuel cell based). The figures indicate that by using a range extender the overall vehicle mass is determined by the electric range. In the illustrated case a (battery-) electric range of 80 km is assumed leading to a mass increase of 200 kg in comparison to conventional cars. At Austrian conditions this range would be sufficient to drive more than 80 % of the annual driving distance in electric mode (see chapter 2.5), while the car is still able to drive longer distances in range extender mode if necessary.

5.5 Economic Assessment of Electrified Propulsion Systems 2010

To diffusion into the mass market a new technology has to meet both technical and economic criteria. This chapter will focus on the economic perspectives of electrification of passenger car propulsion systems.

Firstly a detailed economic assessment of different powertrain systems is performed including all relevant types of costs. The main focus is on investment costs and fuel costs, since they are the most important cost type in particular when it comes to comparing different propulsion technologies. The analysis gives a detailed overview on the technological and economic status of electrified powertrain systems today (2010). To assess the effects of potential cost reductions in key vehicle components and the effect of increasing fuel prices on the overall cost, sensitivity analysis are performed. The results give an impression what framework conditions, in terms of battery costs and fuel prices, are necessary for electric powertrain systems to become cost effective.

Finally, a cost estimation of different powertrain systems in the time frame 2010 - 2050 is made, considering technological learning effects and fossil fuel price scenarios.

Parts of the techno-economic assessments were already presented in previous publications (see (Kloess et al. 2009) and (Kloess et al. 2009b))

5.5.1 Comparing Propulsion Technologies – Reference Vehicles

In order to make the different technologies comparable reference specification are defined for each vehicle class. All vehicles have to offer the same usability in terms of interior space, transport capacity and comfort. Also their driving dynamics has to be comparable. This means that the overall system power has to be the same for all technologies in one class. Consistently, if the user value of all vehicles should be the same, overall driving range would have to be the same for all as well. Due to technical constraints it is not yet possible to achieve comparable driving ranges of conventional cars with battery electric cars. This disadvantage in usability is not reflected in the specific service costs of the vehicles and therefore, has to be kept in mind in the evaluation of the results.

5.5.1.1   Compact class cars:

This class represents small and compact cars. The reference curb weight of this class is 1000 kg and the reference system power is 50 kW. As showed in chapter 2.3 there has been a trend towards higher curb weights in all vehicle classes in the last years and decades. Three decades ago a car that had the characteristics defined here as compact class would have been considered as a middle class car (see chapter 2.4.2). The main drivers of this development are the higher customer expectations concerning comfort, security and driving dynamics even for small cars. Actually, many models that are marketed as compact class cars today have higher weight and power than the vehicle defined here. In 2008 around 20 % of vehicles sold in Austria can be attributed to the compact class as it is defined here (see (Pötscher 2009)).

In this vehicle class only conventional drive systems, micro hybrids and battery electric propulsion systems are analysed. Complex hybrid systems are not considered in this segment due to technical and economic constraints. At these low curb weights the achievable fuel savings are simply too low to justify complex hybrid powertrain systems both from an energetic and economic perspective. In the compact class it makes more sense to switch directly to fully electric cars. Battery electric vehicles are generally a promising option in this class as they are the ideal system for urban short distance trips which is the typical field of application for these types of cars. Furthermore, the reduced driving range would not be critical as in other classes.

Table 5-3: Specifications of Compact Class Vehicles (Data Source: ELEK-TRA-Project (Kloess et al. 2009))

	curb weight	propulsion system		battery (traction)			fuel consumption		driving range	
		engine	electric motor	nominal capacity	ΔDOD (useble cap.)	mass	electricity	fuels	electric	total
	[kg]	[kW]	[kW]	[kWh]	[kWh]	[kg]	[kWh]	[l; kg]	[km]	[km]
Compact Class										
Conventional Drive - gasoline	955	50	-	-	-	-	-	6,0	-	500
Conventional Drive - diesel	989	50	-	-	-	-	-	4,7	-	500
Conventional Drive - CNG	1016	50	-	-	-	-	-	4,2	-	500
Micro Hybrid - gasoline	967	50	-	-	-	-	-	5,4	-	500
BEV 50km + REX (gasoline)	1050	18	50	16	9,6	160	19,7	4,4	49	500
BEV	1037	-	50	24	14,4	240	19,7	-	73	73

The specifications in Table 5-3 show the slight variations of vehicle curb weight for different propulsion systems due to their specific characteristics. It is noticeable that the BEV is only slightly heavier than the other options. The weight of the 24 kWh battery system is compensated through the absence of the entire internal combustion engine, transmission and exhaust system. The driving range of the BEV is considerably smaller than for the other systems. In addition to the pure battery electric car an electric car with lower electric driving range (ca. 50 km with a 16 kWh battery) and a range extender (REX) is analysed.

5.5.1.2 Middle class cars

The reference curb weight of the medium size vehicles is set at 1500 kg and the reference power at 75 kW. Medium sized cars are by far the most important segment of the passenger car market in Austrian (Pötscher 2009). According to this definition, around 70 % of car sales in Austria are medium sized cars. In this category a broad range of propulsion systems is analysed.

Table 5-4: Specifications of middle class vehicles (Data Source: ELEK-TRA-Project (Kloess et al. 2009))

	curb weigth	propulsion system		battery (traction)			fuel consumption		driving range	
		engine / fuel cell	electric motor	nominal capacity	ΔDOD (useble capacity)	mass	electricity	fuels	electric	total
	[kg]	[kW]	[kW]	[kWh]	[kWh]	[kg]	[kWh]	[l; kg]	[km]	[km]
Middle Class										
Conventional Drive - gasoline	1470	75	-	-	-	-	-	7,5	-	700
Conventional Drive - diesel	1522	76	-	-	-	-	-	6,0	-	700
Conventional Drive - CNG	1533	77	-	-	-	-	-	5,2	-	700
Micro Hybrid - gasoline	1495	78	-	-	-	-	-	6,9	-	700
Mild HEV - parallel	1535	65	20	1	-	20	-	6,4	-	700
Full HEV - power split - gasoline	1593	50	50	2	-	30	-	5,9	-	700
PHEV power split - 40km - gasoline	1723	50	50	16	9,6	160	22,2*	5,9	43	700
PHEV series - 40km - gasoline	1608	50	75	16	9,6	160	22,2*	5,5	43	700
BEV 65km + REX - gasoline	1565	30	75	24	14,4	240	22,2*	5,5	65	700
BEV 130km	1692	-	75	48	28,8	480	22,2	-	130	130
FC PHEV 40km - H2	1784	40	75	16	9,6	160	22,2*	0,9	43	500
FCV	1860	80	75	2	-	30	-	0,9	500	500

* value taken from battery electric car

As indicated by the specifications in Table 5-4 curb weight of cars increases with the degree of electrification. The weight growth caused by additional components of a hybrid powertrain, like batteries, motors, controllers and transmission upgrades cannot be compensated by the weight savings through the smaller downsized ICE. Plug-In Hybrids and electric cars have a significantly higher curb weights as a consequence of their heavy battery systems. For example a plug-In hybrid (PHEV) with an electric range of 40 km requires a battery mass of 160 kg when Li-Ion technology is applied. In the battery electric car a 48 kWh battery is used permitting an an electric driving range of 130 km which is raising the curb weight to almost 1700 kg (see chapter 5.4.7).

5.5.1.3 Large vehicles

Large vehicles are defined by a reference mass of 2000 kg and a reference power of 120 kW. This category represents less than 10 % of the vehicle sales in Austria. Only conventional drive systems, micro hybrids, mild hybrids and full hybrids are analysed for this segment.

Table 5-5: Specifications of upper class vehicles (Data Source: ELEK-TRA-Project (Kloess et al. 2009))

	curb weight	propulsion system		battery (traction)			fuel consumption /100km (TTW)		driving range	
		engine	electric motor	nominal capacity	ΔDOD (useable cap.)	mass	electricity	fuels	electric	total
	[kg]	[kW]	[kW]	[kWh]	[kWh]	[kg]	[kWh]	[l; kg]	[km]	[km]
Upper Class										
Conventional Drive - gasoline	2068	120	-	-	-	-	-	9,5	-	700
Conventional Drive - diesel	2151	120	-	-	-	-	-	7,5	-	700
Conventional Drive - CNG	2138	120	-	-	-	-	-	6,7	-	700
Micro Hybrid - gasoline	2093	120	-	-	-	-	-	8,9	-	700
Mild HEV - parallel	2123	100	50	1	-	-	-	8,2	-	700
Full HEV - power split - gasoline	2141	75	75	2	-	-	-	7,7	-	700

5.5.2 Investment Costs 2010

Investment costs are an important factor for the competitiveness of vehicle powertrain systems. Especially for alternative propulsion technologies they have often been a barrier in the past. To assess the investment costs of the different powertrain systems for the status 2010 a component based analysis is performed similar to the one described by (Wietschel & Dollinger 2008) or (IER 2009). In the derived cost model all powertrain systems defined in chapter 5.5.1 are captured.

To identify the main cost drivers for electrified vehicles today (status 2010), the cars are broken down into main component groups:

o **Vehicle basis:** in the analysis it is assumed that all vehicles share the same vehicle basis. It includes all components that are not relevant for the propulsion: the chassis, the undercarriage (including the steering), the interior equipment including all comfort and security features, the exterior equipment (e.g. tyres, mirrors, windows etc.), the entire on board electricity grid (12V) and all the equipment for the control of vehicle functions.

- **Internal combustion engine:** the component group internal combustion engine includes apart from the engine itself, the transmission and the driveshaft
- **Electric drive system:** the electric drive system includes all electric machines, the motor control unit, current converters and the high voltage grid (e.g. 120V)
- **Batteries:** This component group includes both starter and traction batteries. As starter battery ordinary lead acid technology is chosen while the traction batteries are Lithium Ion based cells. Furthermore, the component group includes the battery control units and the thermal control system.
- **Fuel Cell System:** the fuel cell system considered in this analysis runs on compressed hydrogen and includes apart from the fuel cell stacks (PEM Fuel Cell) all necessary auxiliaries such as the thermal management systems, air compressors, DC/DC converters etc.
- **Tank System:** The tank system includes all components for on board storage of liquid and gaseous fuels. The range of systems extends from plastic tanks for conventional fuels to pressure bottles made of composite materials for high-pressure hydrogen storage (e.g. 700bar).

5.5.2.1 Specific component costs 2010

Vehicle Basis: The average cost of the vehicle basis strongly depends on the comfort and safety features the car is equipped with, which also depend on the market that is analysed. For this specific analysis car sales data of Austria was used to define the cost of the average vehicle basis for each of the three defined vehicle classes. (see (Statistics Austria 2009b) & (Autorevue 2010))

For the estimation of the future cost development it is assumed that real cost of the basis would remain the same. This assumption is based on the experience made in the last three decades where the real cost of a vehicle category remained the same even though technological learning effect should have caused cost reductions. This development can be explained by the fact that in the same time the cars became more complex with additional safety and comfort features. This all caused additional cost that outweighed the reductions generated through learning effects.

Internal Combustion Engine ICE: The internal combustion engine is a very mature technology. The basic functioning of the engines has been the same since the beginning of the use of this technology one century ago. During this time period the technology has improved continuously. The engines today have reached a high level in terms of efficiency, emission reduction and power density. Today the engines are close to the physical limitations of the technology defined by the theoretic characteristics of the thermodynamic process. This makes further improvements increasingly difficult. However, even at this high technological state, further improvements of the engines are expected for the next years and decades making them cleaner and more efficient, but also more complex.

That is why for ICEs no considerable cost reductions can be expected in the future. The production process and scale is already on an extremely high level making further improvement difficult. Furthermore, the continuous improvement and the increased complexity is expected to outweigh the learning effects of the technology just like it has done in the past. Therefore, in this analysis the ICE has been considered as a component group without possible cost reduction in the time frame 2010-2050.

In the analysis the cost of the ICEs is quantified in cost per unit of engine power (e.g. € kW^{-1}, $ kW^{-1}). The cost estimation of the ICEs is based on European studies on this field using data from (EUCAR et al. 2006) (see Table A-5 in Appendix A).

Electric Drive Systems Electric machines (EM) used for vehicle traction are quite a new development. As mentioned in chapter 4.5 electric machines were used as traction motors for cars in the early years of automobile history and later for small series cars (mostly prototypes). However, large scale use of EMs in cars for traction or traction support under practical driving conditions is quite a new development triggered by the recent trend toward hybrid cars. Today different technologies of electric traction machines are considered for cars and it is not yet clear which technology is going to become the standard (Permanently Magnetized Synchronous Machine – PMSM or asynchronous machines – ASM).

Since they are applied in many fields production experience for electric machines is already high and it is questionable if these systems will experience considerable cost reduction from production experience and up-scaling. In this analysis it is assumed that the cost of EMs are not going to decrease due to learning effects and will remain constant for the same reasons like ICEs.

For the cost of electric machines in the analysis estimations from literature are used that are based on high production scales (see Table A-6 in Appendix A) (cf. (EUCAR et al. 2006)).

The integration of electric machines in the powertrain requires additional clutches and upgrades of the transmission. This increase in complexity is also reflected in the cost of hybrid systems. To capture this cost effect in the analysis the necessary technical measures are summed up as "vehicle powertrain upgrade" and their costs have been estimated. For the full hybrid (power split) upgrade the cost given by (EUCAR et al. 2006) is used. Due to the lower

system complexity costs of mild hybrid powertrain upgrade is estimated to be only 50 % of the full hybrid upgrade (see Table A-7 in Appendix A).
Another component that has also been allocated to the electric drive component group is the electric charger. For the electric on-board charger additional costs of 500 € have been assumed which is based on the estimations of (Williams & Kurani 2007) who estimated the cost of a charger to be 690 $ (see Table A-7 in Appendix A).

Battery System: today the battery is the main cost driver of electric automotive propulsion systems and will play a key role for their future success. Batteries have been used for more than a century for different application fields. However, the requirements of an automotive traction battery are significantly higher than in most other fields they are used. Especially the requirements in energy and power density, cycle stability and durability (10 years) are asking for advanced battery technology that is especially designed for automotive applications.

Even though there is vast experience in battery development and production for different fields of application, high power and energy traction batteries for automotive applications are quite a new field. This is also reflected in the high cost of batteries that are capable to meet the necessary criteria, like new lithium ion batteries (see Table 5-1). For the future technological progress in battery technology and up-scaling of production is expected to lead to considerable reduction of Li-Ion battery cost (see chapter 5.6).

In this analysis the specific cost of energy batteries was set at 700 € kWh^{-1} for the status 2010 according to (Passier et al. 2007).

Fuel Cell System: Another important component in the context of electric mobility is the fuel cell system. Even though fuel cells have been used for stationary applications for several years, mobile applications are still quite a new field. All cars that were built in the past were experimental vehicles built for fleet testing programs. The fuel cell systems for these prototypes are extremely costly as they are still in an experimental state and are built under conditions that are not comparable to large scale production. This makes it difficult to find reliable estimations on system costs for the 2010 technology status. Most data that is provided by cost analysis for fuel cell systems is based on high production scales of several hundred thousand units per year. For example in the fuel cell cost estimation by (Carlson et al. 2005) a detailed bottom up approach is applied: The entire production process and the applied materials are considered and used as a basis for high production volume cost projections.

The resulting cost potential for the technology status of 2005 is 108 $ kW^{-1}. Table 5-6 gives an overview on cost estimation from different sources.

Table 5-6: Cost estimations for Fuel Cell Systems   Data Source: (EUCAR et al. 2006) (Chalk et al. 2000) (IEA 2008) (IEA 2007) (Kromer & Heywood 2007) (Carlson et al. 2005) (Schoots et al. 2010)

	Projection	Cost [€ kWh-1]	Source
FC-System:	2010	105	EUCAR-CONCAWE-JRC-2007
FC-Reformer:	2010	146	EUCAR-CONCAWE-JRC-2007
FC-System:		120	US DOE 2006
FC-System:	Target 2010	35	US DOE 2006
FC-System:	Target 2020	25	US DOE 2006
FC Stack-controller	Mass Prod.	500	IEA 2008
FC-Stack-controller	Target	100	IEA 2008
PEMFC-Stack	2005	1800	IEA 2007
PEMFC-Stack	2010	500	IEA 2007
PEMFC-Stack	2030 optimitic	35	IEA 2007
PEMFC-Stack	2030 pessimistic	75	IEA 2007
Fuel Cell System	Conservative	75	MIT 2007
	Baseline	50	MIT 2007
FC-System costs:	500k untis/year	97	NREL 2005
FC-System costs:	2007	1067	Schoots et al 2010

First production series will be based on small volumes with high use of manual labour. The cost estimations based on high production scale given by (Carlson et al. 2005) can't be used for the introduction phase (2010-2020) of this technology. In the starting phase production will be based on smaller scales with higher cost per unit.

The analysis done by (Schoots et al. 2010), who determined specific cost of automotive fuel cell systems (80 kW systems) with 1067 € kW^{-1} in 2007, gives a better impression of the 2010 cost level. It is evident that with these high costs fuel cell cars are still much too costly for market introduction. Chapter 5.5.6 will illustrate the necessary specific costs of fuel cells that have to be achieved to make the technology competitive.

Tank System: Since costs of on board storage differ strongly for different fuels, the tank system is regarded as separate component group in the cost model. For conventional powertrain systems using liquid hydrocarbon fuels the cost of

the fuel tank are relatively low. For systems that are using gaseous fuels which have to be stored at high pressures, the tank system becomes an important cost factor (see Table A-8 in Appendix A).
Concerning the cost development assumptions no cost reduction can be expected for conventional tanks. Even the cost of CNG is not expected to reduce significantly since they are based on state of the art technology (steel pressure bottles). The only tank system that is likely to experience cost reduction are the high pressure hydrogen tanks. They are made of composite materials and are very costly today. In the future up-scaling of production is expected to lead to a significant cost reduction of these components.

5.5.2.2   Net Investment Cost 2010

In the cost model the cost of the components ad up to the net investment costs of the cars:

$$IC_{car} = \sum IC_{comp} \quad [€] \tag{5-2}$$

IC_{car} ... net investment costs of the car
IC_{comp} ... net investment costs of the components

Figure 5-26 gives an overview on the net investment costs of the analysed propulsion systems for the status 2010 and illustrates how the different component groups contribute to the overall cost of the cars. It shows that investment costs increase with the degree of electrification, which is mainly driven by the battery costs.

Because of the battery the cost of a BEV with an electric driving range of 130 km is about three times higher than of a CD gasoline car. Plug-In hybrids with 40 km electric driving range still cost about twice as much as conventional cars in this segment.

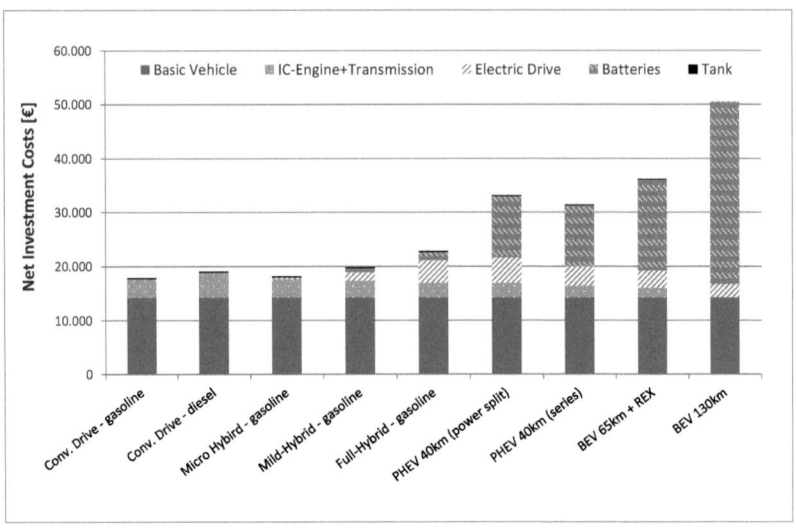

Figure 5-26: Net Investment Cost of Powertrain Systems in 2010 (Middle Class)

5.5.3 Fuel Costs

Fuel costs of vehicles are determined by their fuel consumption and the fuel price. Fuel consumption depends on the efficiency of the car (see chapter 5.3). The fuel price is affected by the net fuel price and different types of taxes (see chapter 4.6.4). The specific fuel costs of the cars in the class i with the technology j and the fuel h are calculated as follows:

$$FC_{ij} = EC_{ij} \cdot FP_h \qquad [\text{€ km}^{-1}] \qquad (5\text{-}3)$$

FC ... fuel costs
FP ... gross fuel price
EC ... energy consumption of the car

5.5.4 Cost comparison of propulsion technologies

From the capital costs and the fuel costs of the cars the specific service costs are derived that can either be expressed in € km^{-1} or € year^{-1} (cf. chapter 3.8.2).

$$SC_{net\_ij} = CC_{SP\_ij} + FC_{ij} \quad \text{[€ km}^{-1}\text{]} \tag{5-4}$$

SC ... service costs [€ km^{-1}]
FC ... fuel costs [€ km^{-1}]
CC$_{SP\_j}$ ... specific capital costs of vehicle j [€km^{-1}]

Investment costs of cars are levelized over the depreciation period of 10 years at an interest rate of 5% using the capital recovery factor.

$$CC_{SP\_j} = (CRF \cdot IC_{ij}) \cdot D^{-1} \quad \text{[€ km}^{-1}\text{]} \tag{5-5}$$

$$CRF = \frac{r \cdot (1+r)^{DT}}{(1+r)^{DT} - 1} \tag{5-6}$$

r ... Interest rate
DT ... Depreciation time
CRF ... capital recovery factor
IC ... Investment costs of vehicles
D ... driven kilometres by year

In the cost assessment net cost of vehicles and gross price of fuels are considered. Other costs like non fuel operational costs (see chapter 5.8) that also contribute to the total cost of ownership are not considered. They are assumed to be the same for all systems (ceteris paribus). Also taxes on cars, like tax on acquisition and tax on ownership (cf. chapter 2.6) are not considered since country specific support schemes of certain technologies would distort the assessment. The adoption of new vehicle propulsion technologies is a process that can only happen on a global or at least supra-national level since only with large scale adoption the cost of the technology can be reduced sufficiently to make it competitive (see chapter 4.3). Even though fiscal instruments or other incentives could favor a technology in one country, the situation might be completely different in another. For a neutral analysis of the economic potential of propulsion technologies the net cost of the technologies have to be analyzed before analyzing one particular country with its specific taxation scheme. Even within the European Union the taxation schemes for passenger cars vary significantly (see chapter 2.6).

In principle there are also differences in fuel taxation leading to dissimilar price levels within the EU. However, the fuel tax in most countries lies within a margin of only +/- 20 % (see Figure 2-17 in chapter 2.6). Furthermore, using only net fuel price would distort the results considerably since the fraction of taxes in the motor fuel prices is relatively high. That is why net car prices are combined with gross fuel prices in the assessment.

Table 5-7: Sensitivity analysis assumptions

depreciation time	10	[years]
interest rate	5	[%]
Fuel Prices 2010		
Gasoline	1.2	[€/l]
Diesel	1.1	[€/l]
Electricity	0.2	[€/kWh]
specific battery cost (status 2010)	700	[€/kWh]

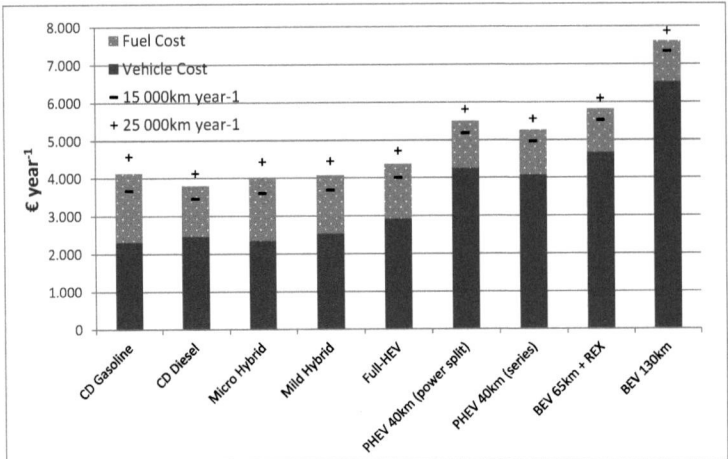

Figure 5-27: Yearly costs at 20 000 km year^{-1} – status 2010 (net vehicle cost & gross fuel price)

Figure 5-27 shows the comparison of net cost (status 2010) of middle class cars with different propulsion systems assuming a yearly kilometrage of 20 000 km, a depreciation time of 10 years and an interest rate of 5 %. 20 000 km is quite a high yearly kilometrage for European conditions. In Austria average kilometrage is around 13 500 km year^{-1} (see chapter Figure 2-5). However, more efficient and therefore more costly technology is usually used by intense users (e.g.

diesel cars have significantly higher average kilometrage than less efficient gasoline cars). For this reason electrified propulsion technology would rather be adopted by intense users with yearly driving distances of more than 15 000 km. The total cost comparison given in Figure 5-27 shows that under the given conditions mild hybrids have about the same cost as conventional cars with gasoline or diesel engines. Complex hybrid systems like the full hybrids with power-split drive are not cost efficient at the given conditions due to their high investment costs. Also pure electric drive systems (PHEVs and BEVs) are not cost effective today. It is evident that yearly kilometrage of the vehicles affects the economic assessment of vehicle propulsion systems. Figure 5-27 also illustrates the sensitivity of total cost with respect to yearly kilometrage.

5.5.5 Sensitivity Analysis

Sensitivity analyses are conducted to assess the impact of increasing fossil fuel prices and the reduction of specific battery costs on the competitiveness of electrified powertrain systems. Figure 5-28 shows the gasoline price sensitivity of the analysed propulsion technologies at the technology cost level of 2010 with specific battery costs of 700 € kWh^{-1} (see chapter 5.2.2.2)[1].

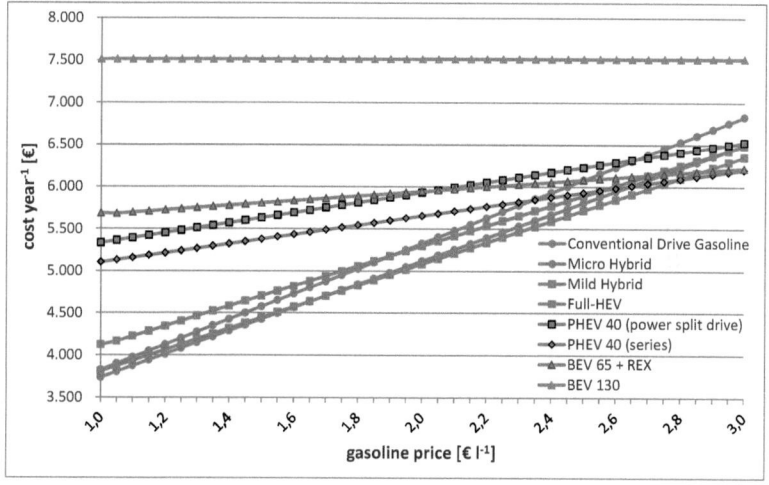

---

[1] The sensitivity analysis on gasoline price is performed ceteris paribus, which means that also the electricity cost would remain constant. In practice higher fossil fuel prices would also affect the electricity price (and even vehicle cost), however the impact would be much lower which makes this comparison admissible.

Figure 5-28: Gasoline Price Sensitivity of propulsions Systems at technology cost status 2010 (spec. battery cost = 700 € kWh^{-1}; 20 000 km year^{-1})

There are three main findings that can be drawn from the analysis:
- Micro and mild hybridisation is cost effective even at the present fuel price levels at a yearly driving distance of 20 000 km.
- Mild hybrids become cost effective at gasoline prices of around 1.6 € l^{-1}.
- At fuel prices above 2.7 € l^{-1} the PHEV-40 with series drive becomes the least cost option. Based on the average Austrian user pattern (see chapter 2.5), it is assumed that an electric driving range of 40 km allows to drive 50 % of yearly kilometrage in electric mode.
- At the present costs status Battery Electric Vehicles BEV are not cost effective, not even at significantly higher gasoline prices.

Since the results show that with specific battery cost of 700 € kW^{-1} (status 2010) electrified propulsion technologies require much higher fuel prices, this raises the question how reductions of specific battery costs affect their cost effectiveness. To find out the necessary cost level a sensitivity analysis with respect to specific battery cost is conducted. Figure 5-29 illustrated the results assuming a gasoline price of 1.2 € l^{-1} and a constant electricity price (see Table 5-7)

The results show that hybridisation is about to become cost effective at high yearly driving distances (20 000 km year^{-1}). Within the different types of hybrid systems micro and mild hybridisation have lower total cost than complex power split full hybrid systems. It is evident that hybrid systems with small batteries show a lower sensitivity to the battery cost than systems with higher electric range and corresponding bigger batteries. A more detailed view on the cost effectiveness of hybridisation measures is taken in chapter 5.5.7.

Pure electric systems like PHEVs and BEVs will need a decrease of battery cost together with an increase of fuel prices to become economically competitive. At the present fuel price specific battery cost would have to be lower than 150 € kWh^{-1} for electric cars to become cost effective.

The corresponding sensitivity analysis for a yearly driving distance of 15 000 km year^{-1} is given in Appendix A.

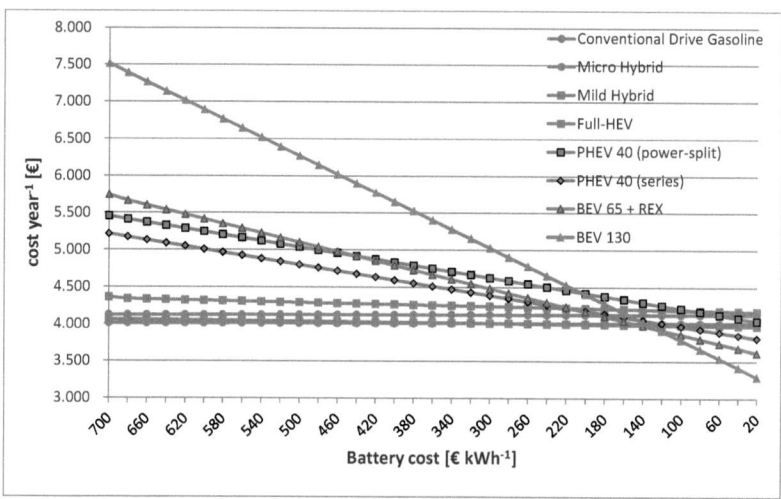

Figure 5-29: Battery Cost Sensitivity of Propulsion Systems at a gasoline price of 1.2 € litre^{-1} (20 000 km year^{-1})

5.5.6 Economic assessment of fuel cell systems

Fuel cell based propulsion systems are often seen as a promising option for zero emission passenger car transport in a long-term period. As already addressed in chapter 5.4.6 the high cost of fuel cell technology is still a major barrier. To successfully enter the market fuel cell vehicles (FCV) will have to be economically competitive with other partly or fully zero emission technologies. Their direct competitors will be battery electric cars and PHEVs. Even though FCVs could have technical advantages (e.g. higher electric range) that could justify a higher price, their costs will still have to be reduced considerably to address a mass market.

Figure 5-30 illustrates how the cost of the fuel cell system affects the net cost of a fuel cell car. Two ways to apply a fuel cell system in the car are analyzed: one with a 40 kW fuel cell system serving as a range extender and another with an 80 kW fuel cell for a pure hydrogen-based fuel cell car. At the present fuel cell cost status (2010) which is approximately 1000 € kW^{-1} (see Table 5-6) a fuel cell car would cost about twice as much as a battery electric car with 130 km range. This underlines that cost reductions of the fuel cell system will be necessary.

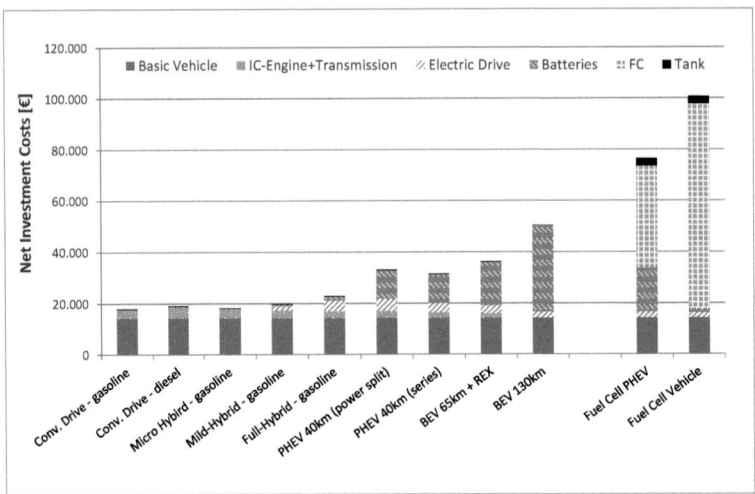

Figure 5-30: Net investment costs of fuel cell cars compared to other propulsion techologies in 2010 (Middle Class)

To find out the necessary specific cost for fuel cell systems to be able to compete with other zero emission technologies a sensitivity analysis is conducted. Hereby, capital costs and fuel costs are considered. Generally, the hydrogen price represents another uncertainty in the economic assessment of fuel cell cars. Estimations of future hydrogen prices for fuel cell cars show a high bandwidth depending on the production pathway. The minimum productuion cost that could be achieved with current technology status and energy prices are estimated between 2 and 3 €/kg for natural gas based production pathways (see (Wietschel et al. 2006) (Altmann et al. 2004) (Ajanovic 2008)). However, the derived consumer price would be considerably higher because of retail margins and possibly also taxes or charges. In this analysis two cases of hydrogen retail prices are assumed: 3 € kg^{-1} (0.09 € kWh^{-1}) and 5 € kg^{-1} (0.152 € kWh^{-1}). Furthermore, the analysis is conducted at two levels of specific battery costs. Figure 5-31 shows that if specific battery costs are 500 € kWh^{-1}, fuel cell cost has to be below 300 € kW^{-1} to undercut battery electric cars and below 150 € kW^{-1} for PHEVs. Lower hydrogen prices thereby permit higher fuel cell costs, but their effect is much smaller because of the high efficiency of fuel cell cars.

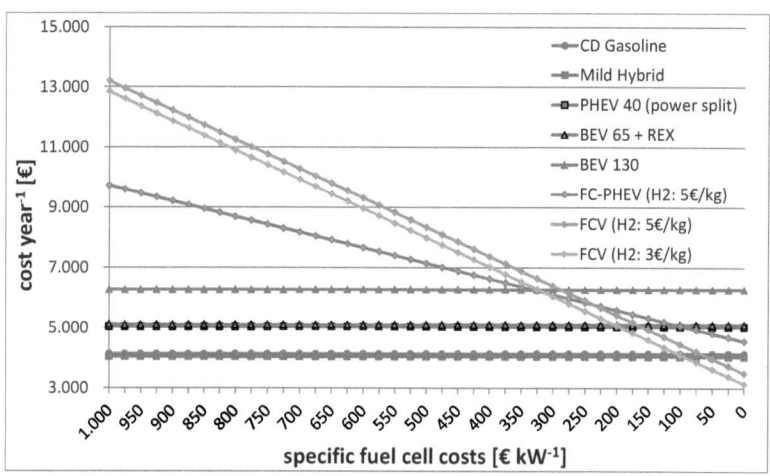

Figure 5-31: Sensitivity of the yearly cost with respect to specific fuel cell costs (Battery Cost = 500 € kWh^{-1}; 20 000km year^{-1})

Figure 5-32 illustrates the sensitivity analysis with battery costs of 250 € kWh^{-1}. At this battery cost level the fuel cell cost has to be below 150 € kW^{-1} for fuel cell cars to be able to compete with a BEV 130 and below 100 € kW^{-1} for PHEV 40. The sensitivity analysis underlines the strong connection of fuel cell cost requirements and battery costs and the fact that future success of these two energy storage approaches will strongly depend on their future cost development.

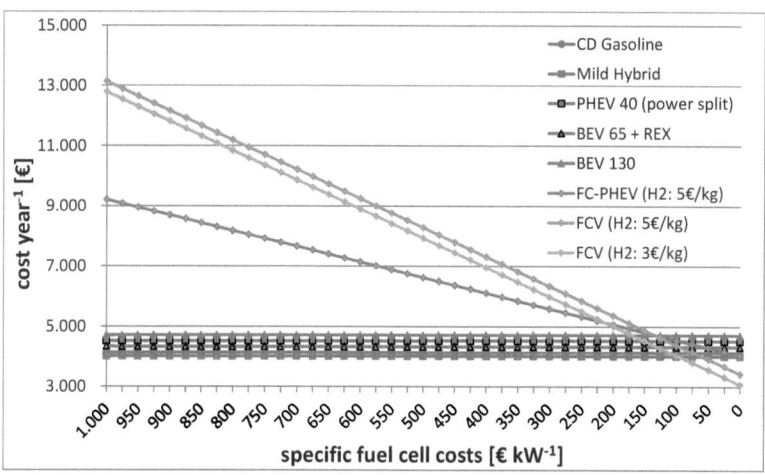

Figure 5-32: Sensitivity of the yearly cost with respect to specific fuel cell costs (Battery Cost = 250 € kWh^{-1}; 20 000km year^{-1})

5.5.7 Net present value of fuel cost savings through hybridisation

From an energy-economic perspective electrification/hybridisation of passenger car propulsion systems can be seen as simple efficiency improvement measure which has to be cost effective to address a mass market. The cost effectiveness thereby strongly depends on the given economic framework conditions. Figure 5-33 compares the net present values (NPV) of fuel savings that can be achieved through hybridisation measures with their estimated costs at different yearly driving distances at economic framework conditions of 2010 (gasoline price: 1.2 € l^{-1}, specific battery costs 700 € kWh^{-1}, hybrid component costs see Table A-7 in Appendix A). The net present value of the fuel savings generated during the vehicle depreciation time indicate what extra additional costs are economically justified at the given fuel price levels. As reference technology the conventional ICE based propulsion systems is used.

$$\Delta IC_{estim\_j} = \sum IC_{comp\_j} - IC_{CD} \quad [\text{€ km}^{-1}] \tag{5-7}$$

$$NPV_{FC\_j} = (FC_{CD} - FC_j) \cdot CRF^{-1} \quad [\text{€ km}^{-1}] \tag{5-8}$$

$$CRF = \frac{r \cdot (1+r)^{DT}}{(1+r)^{DT} - 1} \tag{5-9}$$

ΔIC_{estim} ... estimated additional cost of hybridisation [€]
NPV_{FC_j} ... net present value of fuel saving through hybridisation [€]
IC_{comp} ... component costs [€]
IC_{CD} ... capital cost of conventional drive car [€]
FC ... annual fuel cost [€]
CRF ... capital recovery factor
r ... interest rate
DT ... depreciation time
IC ... investment costs of vehicles

The results show that at the 2010 gasoline price level only micro hybridisation and at higher yearly driving distances, also mild hybridisation are cost effective. Complex full hybridisation and PHEVs are too costly to compete with conventional systems.

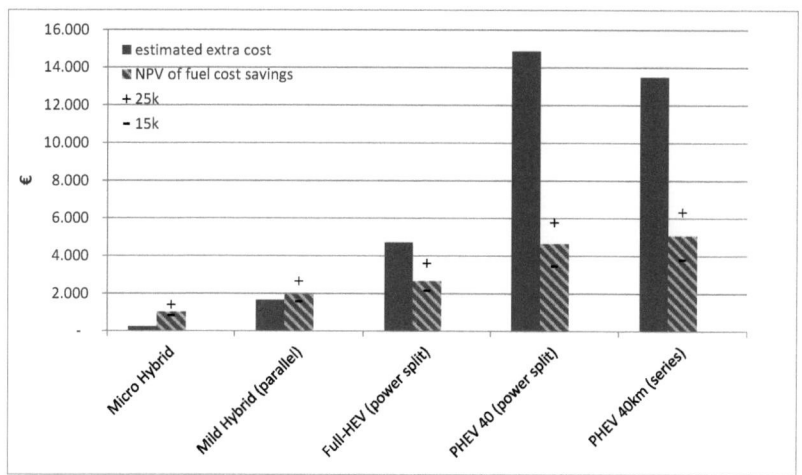

Figure 5-33: Net present values of fuel cost savings versus estimated additional cost of powertrain hybridisation

It is evident that fuel prices have significant impact on the cost effectiveness of hybridisation measures. With increasing fuel price the net present value of fuel savings from hybridisation increases. Figure 5-34 illustrates the effect of the gasoline price on the different hybrid powertrain options at a yearly driving distance of 20 000km.

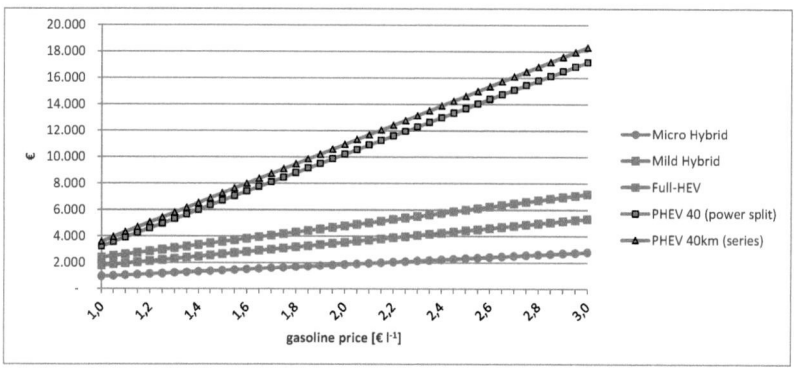

Figure 5-34: Net present value of fuel cost savings through hybridisation as function of the fuel price (20 000 km year^{-1})

5.6 Technological Learning effects of key components

In order to estimate future costs of electrified propulsion systems in the time frame 2010-2050 the concept of technological learning is applied. As explained in chapter 4.3 in the concept of technological learning the cost development is defined by the cumulative production and the learning rate. Since components have different degrees of technological maturity they are expected to experience different learning rates in the future.

As explained in chapter 5.5.2, there are components that are quite new in the field of automotive applications. Especially components which are associated with the recent development of vehicle powertrain electrification are expected to see significant cost reductions in the next decades due to technological learning effects. For other more mature components it is not expected that costs will decrease significantly due to the fact that their cumulative production is already high and additional production would only cause minimal effects. Furthermore, learning effects of mature components are expected to be offset by higher standards (e.g. safety, efficiency, etc.) and the corresponding increase in complexity. Therefore, it is assumed that the internal combustion engine, the electric machines and the vehicle basis are not going to see cost reductions 2010-2050.

5.6.1 Battery Learning Curves

Batteries are the main cost drivers of electrified propulsion systems today, but they are also considered to have high cost reduction potential in the future (see chapter 5.4.4). Costs of batteries are usually defined as cost per unit of energy storage capacity (e.g. € kWh^{-1}). It is expected that the specific cost of batteries will decrease because of higher production experience and production scale. In the meantime, also the technical characteristic of batteries will improve as they did in the past two decades (see chapter 5.4.3). In this analysis the cost development of battery system is estimated through learning effects (theory see chapter 4.3). The estimation is not focused on one specific battery technology, but automotive traction batteries are considered as one component whose cost will decrease as a consequence of learning effects. In the scenario time frame battery cell technologies are likely to change and technical specification will improve. However, these effects are not considered to affect battery cost reduction considerably since technologies will only be adopted when they are economically feasible.

In the technological learning approach cumulative production and the learning rate have to be determined ex-ante which always implicates uncertainties in the estimations of future costs. One possible solution to this problem in an energy model is the use of internalized learning effects, where the modeled cumulative production is affecting the cost reductions. In this case the feedback of cost reductions on the future cumulative production is captured correctly and the learning rate remains the only uncertainty. Since learning effects are usually a global phenomenon, a model would have to cover the global market or at least large shares of it. In the case of global industries like automotive industry it is difficult to apply endogenous learning effects in a model since economic framework conditions vary strongly in different countries and regions.

In this analysis external learning effects are used. To estimate global cumulative production of batteries the concept of technological substitution is applied (cf. chapter 4). It is assumed that today we are at the beginning of a technological substitution process where conventional propulsion systems slowly but steadily get substituted by electrified drive systems (Hybrids & Electric Drives) as the standard propulsion technology for passenger cars. Electric propulsion system hereby will follow the classical S-shaped curve of technological life cycles (see chapter 4.1). Today, the technology is still in an early phase, the so called "introduction" or "childhood phase". The increasing shares of hybrid cars and the emerging of pure electric cars can be seen as indicators for this development (see Figure 5-37). The next step will be the steep growth phase, which will lead to a great shift in automotive development and a significant increase in production volumes of electric drive components and electricity storage systems. This development will be reflected in the development of global cumulative production of these components. In the first years production will focus on hybrid systems. This will create technology spill-overs on components for PHEVs and BEVs and also affect their cost. Based on this global development the growth in market share of electrified cars (hybrid and electric) is estimated (see Figure 5-36 and Figure 5-37). Thereby a growth in global automobile production is assumed as projected by the IEA (IEA 2009b).

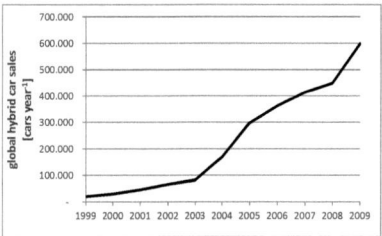

Figure 5-35: Global hybrid sales 1999-2009 (data sources: (US Departement of Energy 2010) (hybridcars.com 2010) (IEA 2009))

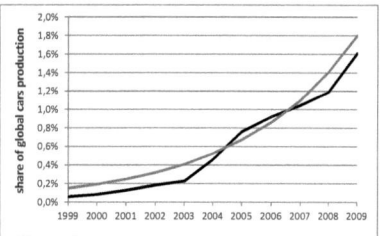

Figure 5-36: Share of hybrids in the global automobile market & fitted S-curve (global car market: (OICA 2010))

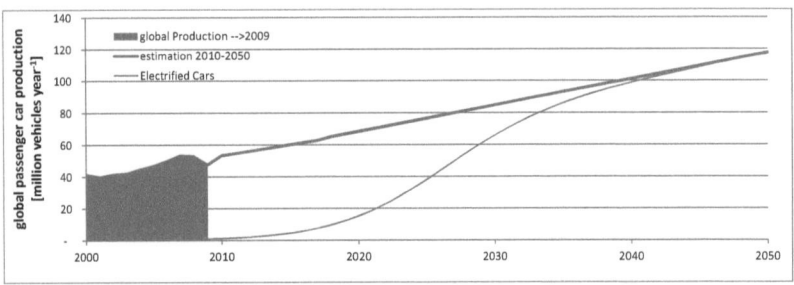

Figure 5-37: Estimates for global production of electrified cars (HEVs & EVs)

It is assumed that in the time frame where electrified cars are substituting for conventional cars also the average battery capacity will increase. Today, passenger cars have power batteries with relatively low battery capacities of around 1 kWh (e.g. Toyota Prius: 1.3kWh; Honda Civic Hybrid: 0.87 kWh). Between 2010 and 2020 however the average battery capacity will also increase due to the stronger diffusion of plug-in hybrids and pure electric cars which will lead to a further acceleration in growth of yearly cell production for traction batteries (see Figure 5-38 and Figure 5-39).

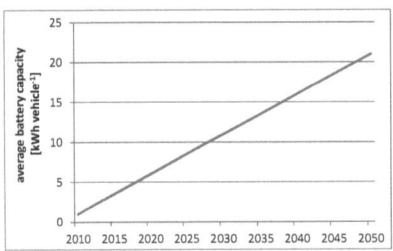

Figure 5-38: Estimated average battery capacity of vehicles

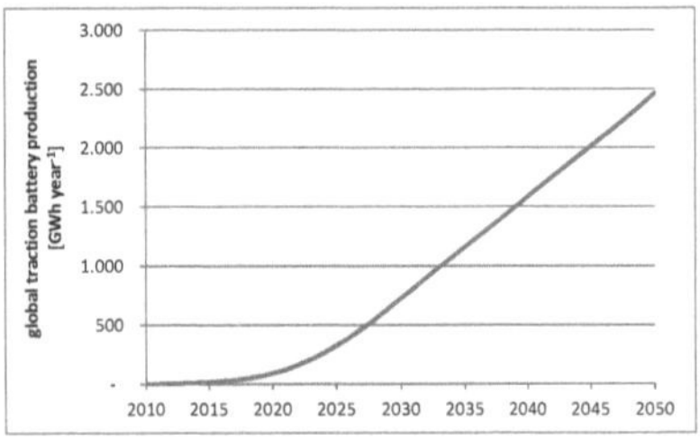

Figure 5-39: Estimated global traction battery production

In the applied approach it is not distinguished between different battery technologies. Today, lithium Ion batteries are seen as the most promising technology for the next decade replacing NiMh batteries in more and more fields. However, it's likely that Li Ion batteries will be replaced by more advanced technologies in the coming decades. Therefore, past learning rates of single battery technologies are not applicable for the analysed time frame 2010-2050 but the learning rate for traction batteries as one technology have to be estimated. (McDonald & Schrattenholzer 2001) shows that the range of learning rates for energy technologies extends from 5 % to 25 %, with an average of around 16-17 %. In the case of traction batteries learning rates ranging from

12.5 % to 17.5 % are used in this analysis. Figure 5-40 illustrates the resulting cost reductions. In the first decades there is a steep cost reductions leading to costs between 212 € kWh^{-1} and 305 € kWh^{-1} in 2020 and 205 €kWh^{-1} and 119€ kWh^{-1} in 2030 (see Figure 5-40).

In the chosen approach the entire cost of batteries experience learning effect, which means that no minimum costs are defined. In other estimations e.g. (Fulton et al. 2009) a fixed part of the cost is defined that represents the minimum costs of a technology which are derived from basic material requirements. The problem with this approach is the fact that defining minimum costs ex ante implicates uncertainties. In the particular case of estimation of future battery cost with a long time horizon (2010-2050) ex ante definition of minimum cost is almost impossible. Battery technologies are likely to change and so will their basic materials. This is why it seemed more appropriate for this particular analysis to apply learning effects on total component costs. Furthermore, minimization of the use of scarce and therefore expensive row materials will always be one development goal for this technology (see next chapter 5.6.2).

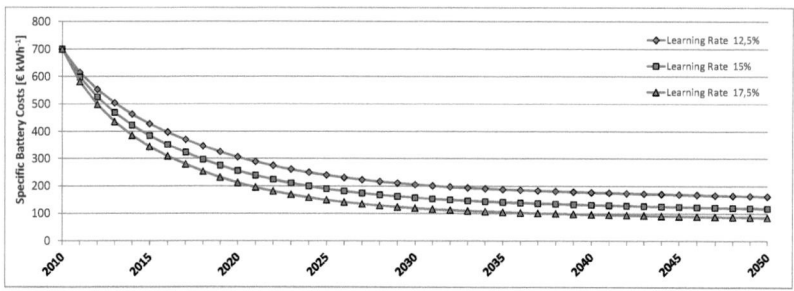

Figure 5-40: Cost Reductions of Batteries with different Learning Rates

5.6.2 Ressource Costraints of key materials

This chapter will briefly explain why scarcity of resources was not considered as a critical barrier to cost reduction of key components of electrified propulsion systems in this work. Since a detailed analysis of this issue would go beyond the scope of this thesis only some of the key facts and arguments are brought up.

Assuming that powertrain electrification will become a global trend, this will inevitably lead to an increasing demand for traction batteries electric machines and other electric drive components. This raises concerns whether this development could lead to a shortage of some key resources for these components. For example the strong focus on lithium Ion batteries today is expected to lead to an increase of global lithium demand (Madani 2009) This raises the question whether scarcity of lithium could become a constraint of electric vehicle diffusion. Scientific analysis of resources for traction batteries sowed that these concerns are not justified. The analysis of (Andersson & Rade 2001) shows that the global resources of lithium are sufficient to build up to 12 billion electric cars with Li Ion batteries. These results are confirmed by other more recent analyses that show that reserves are sufficient to build 1-5 billion cars and resources allow production of up to 10 billion without considering recycling (Schott 2010), respectively that only 49 % of global resources would be exploited in 2050 even in scenarios with high diffusion of electric cars (Angerer et al. 2009b). There is also the concern that growing demand for batteries could cause supply bottlenecks due to limited production capacities for Lithium which could have short term effects on the cost of Li-Ion batteries. The analysis by (Schott 2010) shows that with 1 % of total cost, the share of Lithium raw material cost is very small and even significant price increases would not become a strong cost driver for the cells.

Global diffusion of electrified cars would also drive the demand of other key materials. (Angerer et al. 2010) analyses the future copper demand and availability finding that a considerable increase in copper price can be expected for the next decades because of growing demand. (Angerer et al. 2009a) gives a detailed overview on present and estimated future demand of major resources for future technology. They show that demand is going to increase dramatically in the next two decades for all of the analyzed materials.

However, the shortage of one resource and the resultant price increase usually causes strong efforts to reduce its use. Firstly, there would be a technological development focus on the minimization of specific demand of this particular metal to use it more efficiently, secondly it would drive efforts to substitute it by another metal with better availability, and thirdly the caused price increase would lead to higher recycling rates of used batteries. Furthermore, an increasing price leads to enforced activities to develop these resources.

Consequently, raw material shortages are not seen as a strong driver of electric drive component costs in the future and their impact is not considered in the cost estimation 2010-2050.

5.6.3 Fuel Cell Learning Curves

With a scenario time frame up to 2050 fuel cells have to be considered as technological option. If and when the technology is ever being adopted for automotive purposes remains uncertain. Apart from high costs there are serious technical and infrastructural barriers the technology has to overcome.

To be economically feasible the infrastructure needs an acceptable number of cars on the roads but without the infrastructure the cars will not be (see chapter

4.4). Today there is no transition technology in sight that could pave the way for this technology like HEVs and PHEVs could do for pure EVs. Therefore, mass market introduction (>500 000 units year^{-1}) before 2020 seems unlikely and it's questionable whether the theoretical cost estimations listed in Table 5-6 can be achieved soon.

To overcome the "chicken-egg problem" and to bring down production cost, fuel cell systems will require high production scales even in their introduction phase. Since today (2010) there are no indications that large scale fuel cell production is about to start in the next decade it will be difficult to close the cost gap in a short to mid-term period. Some experts doubt whether it will be possible to close this gap even in a long-term period. According to them potential cost reductions in the next decades could either be generated from technology spill-overs for stationary fuel cell applications or from learning by searching rather than from learning by doing (Schoots et al. 2010).

With this technological breakthrough being the precondition for mass production of automotive fuel cells it is difficult to estimate the schedule of their future cumulative production. Without reliable estimations of global cumulative production for the next decades, learning effect-based cost projections are highly uncertain.

Nevertheless, hydrogen fuel cells are often considered as a long-term option for passenger car propulsion systems. With the long time frame of this analysis (2010-2050) the case of a technology breakthrough of fuel cells has to be considered. For this reason a hypothetical brake through scenario is created assuming that first high volume production (>100 k units year^{-1}) of fuel cell system starts in 2020 with specific costs of 250 € kWh^{-1} a cost where fuel cell systems might be able to address the early adopter market segment. Once introduced in the market, growth in cumulative production will be slower than for electrified since the refuelling infrastructure remains a barrier that will take at least two decades to overcome. Figure 5-41 shows the corresponding diffusion of fuel cell cars in the global passenger car market and Figure 5-42 illustrates the derived yearly fuel cell systems production given that average size of fuel cells systems will increase from 40 kW unit^{-1} in 2020, to 80 kW unit^{-1} in 2050.

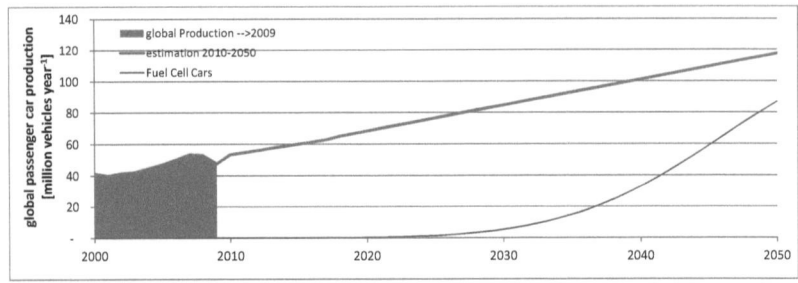

Figure 5-41: Scenario for global fuel cell vehicle production

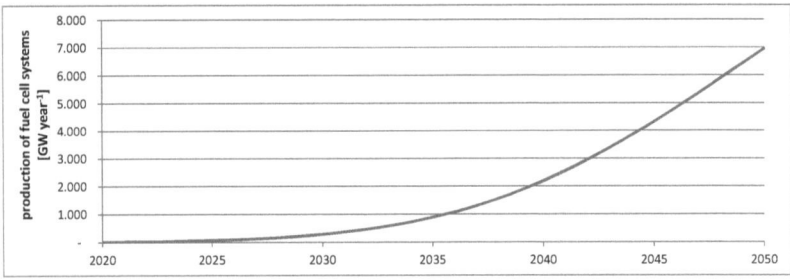

Figure 5-42: Scenario for global fuel cell production

The other critical factor for fuel cells is the learning rate. In past cost estimations for fuel cell systems, learning rates ranging from 10 % to 30 % can be found. However, most simulations use learning rates of around 20 % (Schwoon 2008). In this analysis the same learning rates are used like for battery systems (12.5 %; 15 %; 17.5 %). Figure 5-43 depicts the cost reduction resulting from the assumed cumulative production with different learning rates. According to this estimation fuel cell cost in 2030 will be between 67 € kW^{-1} (LR=17.5 %) and 100 € kW^{-1} (LR=12.5%) in 2030 and between 22 € kW^{-1} (LR=17.5 %) and 46 € kW^{-1} (LR=12.5%) in 2050.

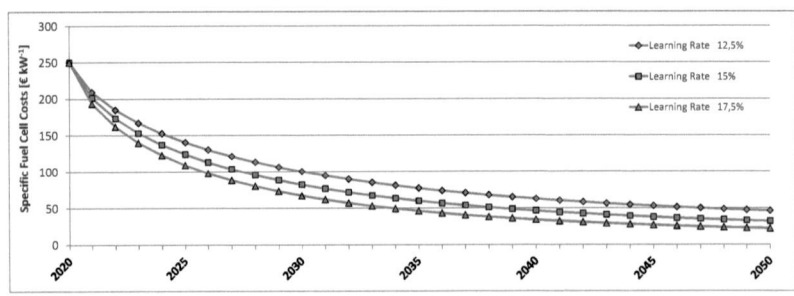

Figure 5-43: Cost Reductions of Fuel Cells with different Learning Rates

These values correspond approximately to the values found by (Tsuchiya & Kobayashi 2004), who made a bottom-up analysis of fuel cell cost reduction potential. They analysed the production processes of the main fuel cell component and their potential cost reduction that can be achieved in large scale production by using learning effects for each of these production processes. The minimum stack cost found was 38 $ kW^{-1} at high progress ratios for every process step and high production volume (5·10^6 vehicle year^{-1}).

Figure 5-44 shows the cost comparison of propulsion systems for middle class cars. Even though learning effects lead to a considerable cost reduction in the time frame 2010-2030, batteries and fuel cells remain strong cost drivers.

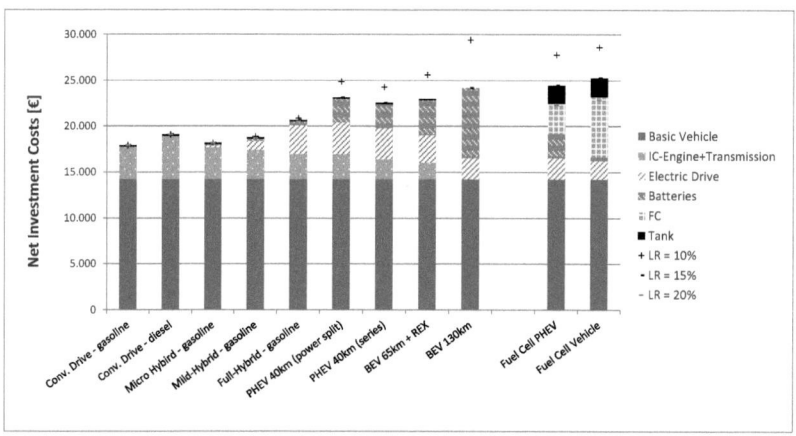

Figure 5-44: Net Investment Cost of Powertrain Systems in 2030 (Middle Class Cars)

5.6.4 Fuel price and Fuel Price Scenarios

Prices of different fuels h are determined by their net prices FP_{net} the fuel tax FT and the value added tax.

$$FP = (FP_{net} + FT) \cdot VAT \quad \text{[€ kWh}^{-1}\text{]} \tag{5-10}$$

FP ... gross fuel price
FP_{net} ... net fuel price
FT ... fuel tax
VAT ... value added tax

The general background of fuel taxation and the fuel tax in Austria have already been discussed in chapter 4.6.4. For the status 2010 in the cost comparison the existing fuel tax level in Austrian (spring 2010) is used and for the net fuel prices status of 2010 current prices in Austria are taken (status: first half 2010). To represent the net fuel price development in the cost scenarios for the time frame 2010-2050 two fossil fuel price scenarios are assumed. A detailed description of the two scenarios will be given in chapter 7.1. Both scenarios show a continuous increase in fuel price up to 2050 with different dimensions of price increase:

Low-Price-Scenario: in this scenario it is assumed that the real fossil fuel price shows a moderate increases that leads to a doubling of fuel prices up to 2050 (75 \$ bbl^{-1} → 150 \$ bbl^{-1}). This assumption is based on the "PRIMES-high" energy price scenario (Kapros et al. 2008).

High-Price-Scenario: in this scenario it is assumed that real fossil fuel prices increase much stronger leading to a tripling of fossil crude oil prices up to 2050, which would mean an oil price of approximately 225 \$ bbl^{-1}.

5.7 Cost Scenarios 2010 – 2050

The results of the economic assessment in chapter 5.5 indicate that cost effectiveness of electrified propulsion systems is mainly affected by the key parameters: fuel price and battery costs.

To estimate the potential of the technologies up to 2050 cost scenarios are developed that are based on the estimations of battery cost and other components (see chapter 5.6) and on global fuel price scenarios (see chapter 7.1).

Based on these assumptions a cost projection is conducted. Figure 5-45 and Figure 5-46 illustrate the corresponding cost developments of the different technologies in the *"Low-Price-Scenario"* respectively in the *"High-Price-Scenario"*, assuming a yearly driving distance of 20 000 km year^{-1}. In the first years the least cost technologies are conventional drive (CD) cars and micro and mild hybrids. In both scenarios battery electric vehicles with range extenders and plug-In hybrids (PHEV & BEV+REX) become cost effective in a mid to long term (starting in 2020 in the "High-Price-Scenario" and in 2030 in the *"Low-Price-Scenario"*). In the *"Low-Price-Scenario"* the differences between all powertrain systems are not significant, with only slight advantages for Hybrids up to 2030 and PHEVs and BEVs later. Due to the strong fuel price increase in the *"High-Price-Scenario"* Plug-In-Hybrid and pure electric propulsion systems become clearly the best options from an economic perspective with savings of several hundred Euros in comparison to convention technologies (see Figure 5-46).

For the corresponding cost estimation for yearly driving distance of 15 000 km year^{-1} see Appendix A.

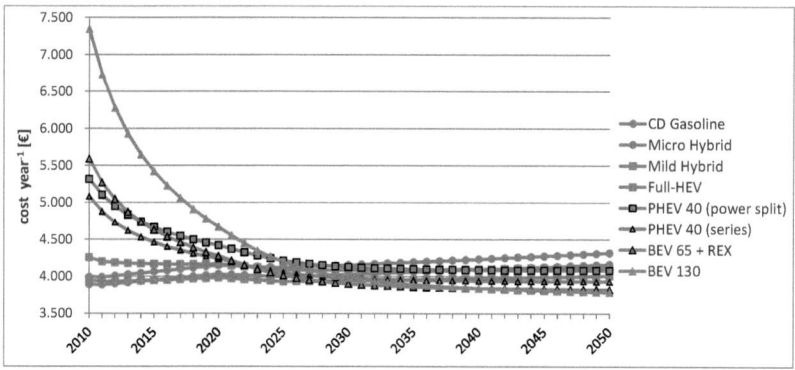

Figure 5-45: Estimated development of yearly costs of propulsion systems in the middle class 2010 – 2050 in the *"Low-Price-Scenario"* (20 000 km year^{-1})

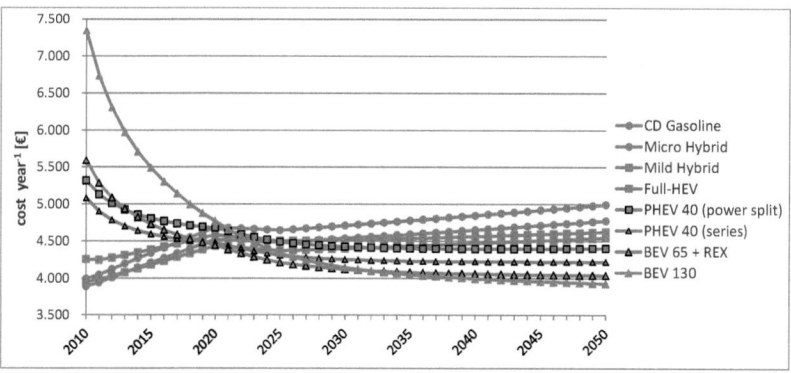

Figure 5-46: Estimated development of yearly costs of propulsion systems in the middle class 2010 – 2050 in the *"High-Price-Scenario"* (20 000 km year^{-1})

Figure 5-47 and Figure 5-48 show the 2030 cost projection for both fuel price scenarios. In the *"Low-Price-Scenario"* all technologies are more or less on the same level. In the "High-Price-Scenario" electric propulsion systems have significantly lower cost than conventional and hybrid systems. Generally the results shows that the higher investment costs of electrified propulsion systems get outweighed by the lower fuel cost because of the better efficiency of electric powertrain systems. The results of the *"High-Price-Scenario"* also indicate that electric powertrain systems are significantly less sensitive to increases in global fossil fuel prices.

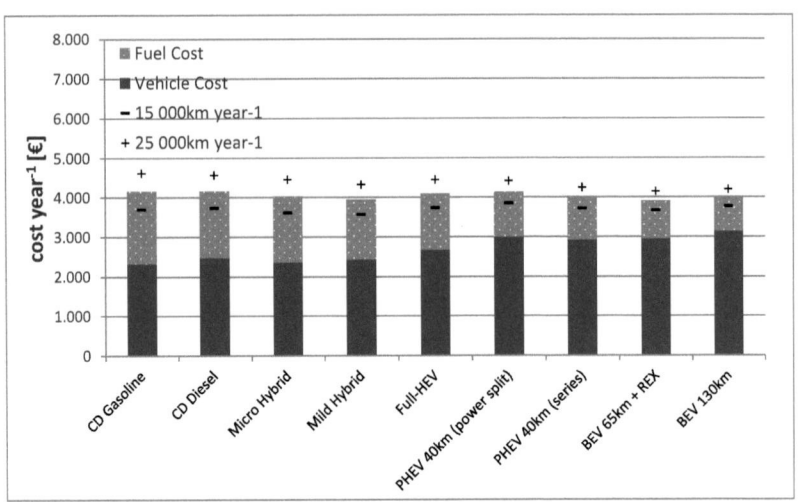

Figure 5-47: Projection of yearly cost in 2030 in the *"Low-Price-Scenario"* (net vehicle cost & gross fuel price)

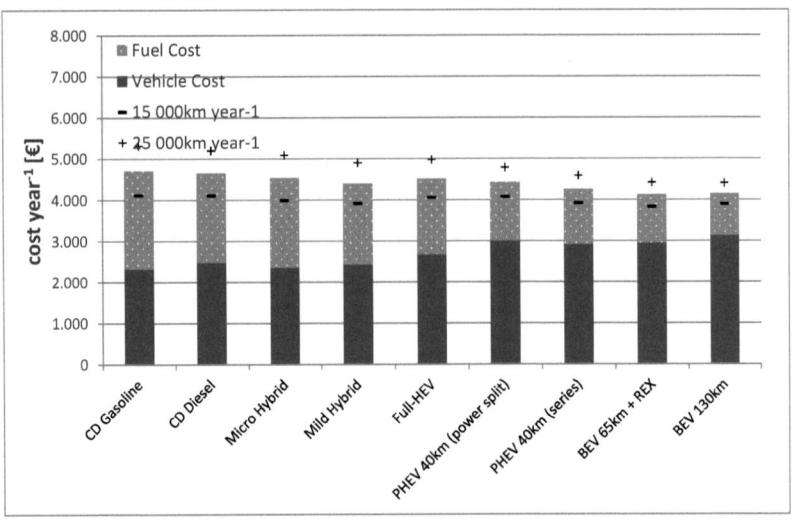

Figure 5-48: Projection of yearly cost in 2030 in the "High-Price-Scenario" (net vehicle cost & gross fuel price)

5.8 Total cost in Austria 2010

The fuel price increases described in the previous chapter were driven by assumed global fossil fuel price development. Since a considerable share of the price of motor fuels usually are taxes (see chapter 2.6.3), they can also have significant effects on the economic competitiveness of propulsion technologies. Another factor that might affect the economy of technologies can be taxes on cars (see chapter 4.6.4).

Hence, for a cost assessment in one specific country or region total cost including fiscal framework conditions have to be considered. In this chapter a country-speicific cost analysis is performed for the case of Austria. The results represent a detailed overview on the costs of passenger car transport for the particular case of Austria in 2010.

In general the specific service costs of each vehicle of the vehicle class i and the vehicle technology j are determined by their specific fuel costs, their specific investment costs, the corresponding taxes and non-fuel operational costs:

$$SC_{ij} = CC_{SP\_ij} + FC_{ij} + OC_{ij} + INS_i + TO_j \quad \text{[€ km}^{-1}\text{]} \tag{5-11}$$

with

SC ... service costs [€ km^{-1}]
FC ... fuel costs [€ km^{-1}]
CC_{SP_j} ... specific capital costs of vehicle j [€ km^{-1}]
OC_{ij} ... non fuel operational costs [€ km^{-1}]
INS_i ... insurance cost [€ km^{-1}]
TO ... tax on ownership [€]

The specific capital costs are strongly affected by different types of taxes. Apart from the value added tax (20%) on both vehicles and fuels there is a tax on acquisition and a taxes on ownership of cars (see chapter 2.6). Specific capital costs are calculated according to the following equation:

$$CC_{SP\_j} = (CRF \cdot (IC_{ij} + TA_{ij}) \cdot (1+VAT)) \cdot S_{km}^{-1} \quad \text{[€ km}^{-1}\text{]} \tag{5-12}$$

r ... interest rate [%]
DT ... depreciation time [years]
CRF ... capital recovery factor
IC ... Investment costs of vehicles [€]
TA ... tax on acquisition [€]
VAT ... value added tax [%]
S_{km} ... kilometres driven per year [km year^{-1}]

There are some other cost categories that have to be considered in a detailed costs assessment. For example costs for insurance and maintenance of the cars. In Austria insurance costs (of the obligatory third party insurance) depend on the engine power of the car which means that they are the same for all propulsion technologies in the analysis (motor own damage insurance would of be different as it also implies the vehicles purchase price). Concerning maintenance costs there are discussions whether they are higher for electrified vehicles or lower during their life time. In most studies maintenance costs for electric cars are considered lower than for conventional cars (see (Werber et al. 2009) & (Turcksin et al. 2008)). What has to be considered is the fact that the lack of persistence of battery systems might drive maintenance costs up during the vehicle lifetime. By now no sufficient data on lifecycle maintenance costs of plug-in hybrids, electric cars and fuel cell cars is available. Therefore, it is assumed that maintenance costs are the same for all propulsion technologies.

Figure 5-49 gives an overview on the total cost of ownership including insurance, maintenance cost and taxes. The comparison with Figure 5-27 points up the significant impact of taxes on total cost of ownership.

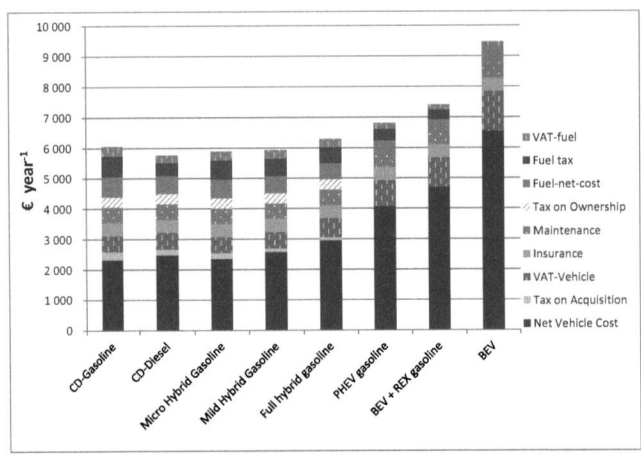

Figure 5-49: Yearly total costs ownership of vehicles at 20 000 km year^{-1} – middle class 2010

6 Simulation model of the passenger car sector in Austria (ELEK-TRA-Model)

In the previous chapter technical and economic aspects of electrified propulsion technologies were discussed. It was showed that HEVs are about to become cost effective in the next years and even PHEVs and BEVs could become cost effective in the short to medium-term. The following chapters will focus on potential effects of large scale diffusion of these technologies on the passenger car fleet. To analyse these effects a simulation model for the Austrian passenger car fleet is used. The model has been developed starting 2007 in the course of two research projects ((Haas et al. 2008), (Kloess et al. 2009)) funded by the Austrian federal ministry of transport, innovation and technology. Some aspects of the model are also described in (Kloess et al. 2010) and (Kloess & Müller 2011).

6.1 Methodology

According to the definition given by (Greening & Bataille 2009), the model represents a bottom-up and top-down hybrid model. It combines a bottom-up model of the Austrian passenger car fleet with top down models that determine passenger transport demand and service level. The bottom-up model includes detailed data on the Austrian passenger car fleet including vehicle specifications, technologies, user behaviour and the resultant energy consumption and energy carriers. To capture long term effects of price and income development, top down models of passenger car transport demand and transport service level are used.

The model can simulate effects of technological development and changing political and economic framework conditions on the passenger car fleet. The impact of changing fossil fuel prices and different fuel- and vehicle taxation schemes on the passenger car fleet in terms of fleet size, vehicle specifications, efficiency, vehicle use and diffusion of technologies can be analysed through scenarios for the time frame 2010-2050. Energy consumption and greenhouse gas emissions are captured on a well-to-wheel basis.

The relatively long scenario time-frame is chosen because of the considerable inertia in the regeneration of the car fleet. Once registered a car usually remains in the fleet about 10 to 15 years. Therefore, it takes decades to see the effects of technological developments on the energy consumption and greenhouse gas emissions. Since potential new technology options cannot be anticipated reliably for such a long time, this long scenario time frame also implicates uncertainties concerning the long-term development of technologies.

One innovation of the model is the detailed coverage of the recent technological trend toward electrified propulsion systems. The model is able to assess the impact of new propulsion technologies on energy consumption and greenhouse gas emissions in the fleet under different political and economic framework conditions. Thereby, it can help to identify the main driving forces for the diffusion of efficient cars and to find optimal policy strategies that support them.

With its special focus on the passenger car fleet the model does not directly capture the impact of other transport modes on the development of passenger car transport. The scenario results have to be seen as ceteris paribus, considering only passenger car transport without any fundamental changes in the price or attractiveness of alternative modes. Unlike other transport models, for example (Zachariadis 2005), (Fulton et al. 2009), or (Ceuster et al. 2007), where the passenger car is one transport mode amongst others in a global model of the transport sector, this model focuses especially on the passenger car fleet, with a strong focus on the specific policies and technologies. Thus, vehicle technologies and political aspects are captured in more detail than in other models.

A further strength of the model is its detailed coverage of the energy supply of the sector including well-to-wheel (WTW) energy and greenhouse gas balances of conventional and alternative conversion chains. This is particularly relevant for future scenarios where a broader variety of energy carriers is expected to be involved.

6.2 Structure of the model

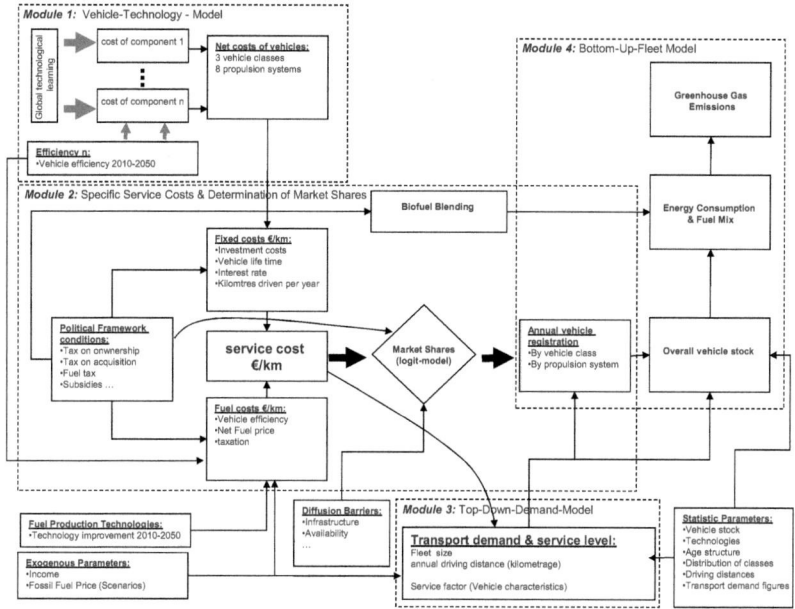

Figure 6-1: Scheme of the model

The global structure of the model is depicted in Figure 6-1. It consists of four main modules.

Module 1: The first module is the vehicle technology model where different vehicle powertrain options are modelled bottom-up to analyse the influence of technological progress on their costs (see chapter 5.5.2).

Module 2: The second module derives market shares of technologies based on their specific service costs considering different levels of willingness to pay. The heterogeneity in consumer preferences is modelled using a logit-model approach with specific service costs as the main parameter (see chapter 6.3). The technology-specific diffusion barriers that arise from limitations in performance characteristics or lack of availability etc. (cf. chapter 4.4) are modelled by predefined constraints of maximal growth in market share of each technology (see chapter 6.3.1).

Module 3: The third module includes the top down models that capture the influence of income, fuel prices and fixed cost on the demand for passenger car transport and transport service level (see chapter 6.4).

Module 4: The fourth module is a bottom-up fleet model of the Austrian passenger car fleet. The fleet is modelled in detail considering age structure, user categories and main specifications of the cars (e.g. engine power, curb weight, propulsion technology, specific fuel consumption, greenhouse gas emissions etc.). The settings are based on a data pool including detailed information about the fleet today and time series of its historic development between 1980 and 2008 (Statistics Austria 2009b) (see chapter 2.3).

6.3 Market shares of technologies

Simulating future market shares of technologies in an energy model always incorporates uncertainties. Consumers are generally defined by inhomogeneous economic framework condition, individual preferences, different levels of willingness-to-pay and asymmetric information, which altogether make it difficult to break the decision-making process down to a simple parameter. To incorporate this inhomogeneity in consumer choice in estimations of market shares in energy models discrete choice models can be applied (Jaccard 2009) (Train 2009).

In this model market shares of propulsion technologies are determined through a multi-nominal discrete choice model with specific service costs of technologies as main decision criteria for consumers. This implies that the purchase decision for a vehicle propulsion system is based mainly on economic criteria if all options offer the same service level.

In the applied approach the discrete choice is implemented by using a multi-nominal logit-model, as described in (Train 2009), (G. Erdmann & Zweifel 2008) and (Axsen et al. 2009). Hereby, the market share z_j of a technology is given by a function of the likelihood w that the technology is been chosen by the consumer on the basis of its specific costs SC_j. Moreover, it is influenced by the specific cost of competing technologies and the reference technology SC_{j_ref}, which is defined as the technology with the largest market share the previous year:

$$z_{j(t)}[\%] = f(r(SC_{j(t)}, SC_{1(t)}, \ldots SC_{n(t)}, SC_{ref(t)}, b), SC_{j(t-1)}, a_j) \quad (6\text{-}1)$$

with

$$r_{j(t)}[\%] = \frac{w(SC_{j(t)}, SC_{ref(t)}, b_s)}{\sum_{j=1}^{k} w(SC_{n(t)}, SC_{ref(t)}, b_s)} \quad (6\text{-}2)$$

and

$$w_{j(t)} = (1 - \frac{1}{(1 + e^{-b_s *(SC_{j(t)} - SC_{ref})})}) \quad (6\text{-}3)$$

z_j ... market share of the technology j [%]
SC_j ... specific service costs of technology [€ km^{-1}]
SC_n ... specific service costs of competing technologies [€ km^{-1}]
SC_{ref} ... specific service costs of the reference technology [€ km^{-1}]
a_j ... diffusion barriers of the technology
b_s ... S-curve parameter
k ... number of technology options

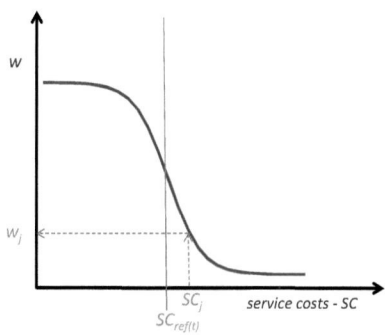

Figure 6-2: Multi-nominal Logit Model

The parameter *b* defines the slope of the logit-function which means that it represents the sensitivity of the market with respect to cost-differences between the options (cf. (Jaccard 2009)). The parameter was determined by using historic data of the Austrian car market (1990-2008). In this time period there was a significant shift from gasoline cars toward diesel cars driven by the cost difference of the two technologies. Figure B-1 in Appendix B shows how the models parameterisation captures the historic development in this time period.

The market share in one year is constrained by diffusion barriers summarized with the variable a_j. They define the maximum growth a technology can achieve with respect to the previous year. (see chapter 6.3.1.)

There are also other approaches future market-shares can be simulated in scenario models. The heterogeneity of customer preferences can also be captured by dividing them in different groups of preferences as demonstrated in (Mock et al. 2009). The main problem with this type of model is gaining sufficient data on consumer preferences to be able to define user groups. Another aspect that needs to be considered is the fact that customer preferences change in the course of time leaving uncertainty for mid- to long-term scenarios.

Therefore, a pure cost-based approach with specific service costs as main decision criteria seemed more appropriate for the particular questions and the time frame of this analysis. Since service costs of cars strongly depend on the annual kilometres and the vehicle category (see chapter 5.5) the car fleet is split up in three vehicle categories and three user groups defined by their yearly kilometrage (see Table 6-1). In the model the market shares of propulsion technologies are determined for each of the resulting segments separately adding up to the cumulative market shares in the entire passenger car market.

$$z_{j\_cum(t)} = \sum_i \sum_u z_{j(u,i)} \cdot z_{u(i)} \cdot z_i \qquad (6\text{-}4)$$

z_{j_cum} ... share of a technology in the entire car market
$z_{j,i,u}$ ... market share of a technology within in a user group u of the vehicle class i
z_{ui} ... share of a user group in the vehicle class
z_i ... share of a vehicle class

6.3.1 Diffusion Barriers

In the modelling of market shares technology-specific diffusion barriers are also considered.

As it was addressed in chapter 4.4 new technologies often have to face barriers that can slow down their market diffusion. In the case of vehicle powertrain technologies, these barriers can have different causes, e.g. an incomplete infrastructure for a certain type of fuel, limited range of models available, limitations in use (e.g. low driving range) etc.

Together, these barriers constrain the potential growth rates in market shares of a technology. This limitation can be represented by the classical s-shaped curves of technological diffusion (Grübler 1998). Past experience with innovation in the automobile industry has shown that their diffusion can vary strongly depending on the technology (S. Jutila & J. Jutila 1986). The shape of the curve used in the model is technology specific, and was determined through evaluation of the specific diffusion barriers for all considered technologies. The diffusion curves for each technology are the upper boundary for growth in market share. Figure 6-3 shows the curves that are applied in the model for different types of technologies. It is assumed that all technologies could theoretically reach 100 % market share. The shortest possible period that a technology needs to penetrate the market (Δt, 10 % - 90 %) is set at 10 years (Technology type A) for technologies which are completely compatible with the infrastructure and meet the expectations of the consumers without any adoption barriers (e.g. micro hybridisation). A historic example for such a technology was the diffusion of downsized diesel engines described in (Cuenot 2009) (see chapter 4.2).

Today, there are propulsion systems which are unable to achieve 100 % market share with their current technological status, since they are not able to meet the requirements of all users (e.g. limited driving range of battery electric cars). However, in the future technologies are expected to improve and consumer expectances could change. This makes it impossible to predefine maximum potentials of one technology in the time frame 2010-2050.

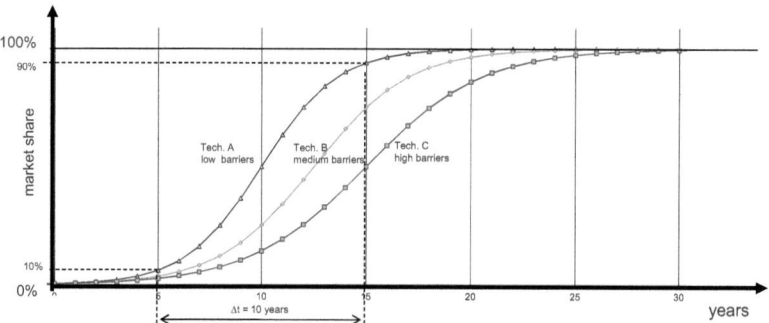

Figure 6-3: Diffusion curves serving as upper boundary for growth in market share of vehicle propulsion technologies within yearly car sales

In the model the growth in market share Δz of a technology j is constrained by upper boundary Δz_D determined by the technology specific diffusion curve:

$$z_{j(t)}[\%] = z_{j(t-1)} + \Delta z_j \qquad (6\text{-}5)$$

with the constraint that

$$\Delta z_{j(t)} \leq \Delta z_{j\_D} \qquad (6\text{-}6)$$

$$\Delta z_{j\_D(t)} = z_{j(t-1)} \cdot \left[\frac{z_{j\_D(t)}}{z_{j\_D(t-1)}} - 1 \right] \qquad (6\text{-}7)$$

$$z_{j\_D(t)} = \frac{1}{1 + e^{a \cdot \{c(z_{j(t-1)}) - b\}}} \qquad (6\text{-}8)$$

Δz_j ... growth in market share of a propulsion technology j
Δz_{j_D} ... upper constraint of market share growth of propulsion technology j
z_{j_D} ... diffusion curve
c_j ... technology specific variable (market share-dependent)
b_j ... technology specific constant

6.4 Modeling the demand and service level of passenger car transport

As explained in chapter 3 total energy demand of passenger car transport is defined by the demand for the transport service, by its efficiency and by the service level (or service quality) (see (Haas et al. 2008)). The coherence can be described in principle by the following equation:

$$E = f(S, \eta(J, F)) \qquad (6\text{-}9)$$

E ... energy demand
S ... service demand (e.g. vehicle kilometres year^{-1})
η ... efficiency of passenger car transport
J ... technologies
F ... service level/service quality (e.g. curb weight & engine power)

As described in chapter 3.2 the demand for a transport mode is determined mainly by its cost (fixed and variable), by the income and by the costs of alternative modes. The transport service level is determined by income and costs of the mode only.
In the chosen approach changes in the costs of other transport modes are not considered (ceteris paribus) and only income and cost of passenger car transport (distinguishing between fixed and variable costs) are considered as determining factors.

In the model the total service demand is expressed by the yearly driving distance of cars (kilometres travelled per year) and by the number of cars in the fleet.

$$S = f(CAP, D) \tag{6-10}$$

CAP ... number of cars in the fleet
D ... distance travelled

Service level or service quality in passenger car transports defines the comfort the transport service is provided and is represented by the average curb weight and engine power of cars.

The following chapters will explain how demand and service effects are captured in the model.

6.4.1 Fleet development

The fleet growth is determined by the income development, expressed by the GDP (an average GDP growth of 1.5% is assumed), the fuel price and the fixed costs. Thus the development of the annual vehicle demand is given by the elasticity of fuel price $α_{FP}$, vehicle price $α_{IC}$ and income $α_y$ (see equation 6-11).

In the model the elasticity of vehicle stock with respect to income is assumed to decrease linear during the scenario time frame from 1.0 in 2010 to 0.2 in 2050. Deriving elasticities for future scenarios always incorporates uncertainties as they are influenced by many factors, e.g. a strong influence of absolute income level on the income elasticity of vehicle ownership. The ratio of car ownership to income growth as well as the income elasticity of vehicle ownership is highly dependent on the country's absolute income level (see chapter 3.4). This correlation can be described by the Gompertz function (J. Dargay et al. 2007). In countries with high income levels like Austria the car fleet is already in saturation and the income elasticity is decreasing. In the model it is assumed that income elasticity would decrease from 1 to 0.2 in the time frame 2010-2050. The starting value for 2010 is derived from calibration runs and is slightly above the range of values found by (Goodwin et al. 2004) who analysed data from UK and comparable countries showing that vehicle stock elasticity with respect to income lies between 0.32 (short term-ST) and 0.81 (long-term-LT) in developed countries.

There are also indicators that price elasticity varies depending on the price level and the direction of price change. (Dreher et al. 1999), who analysed the energy price elasticities of energy-service demand in passenger transport, found that consumers react more sensitive to price changes at higher fuel price levels and also more sensitive to fuel price increases than to decreases. Since no passenger car transport-specific data on this effect is available constant price elasticities are assumed in this analysis.

Calibration of the model based on historic data on the Austrian passenger car fleet (1990-2009) lead to an elasticity of car stock with respect to fuel price of -0.2 and of -0.5 with respect to car purchase price (see Figure B-2 in Appendix B). Both values correspond with the results from the international analysis performed by (Goodwin et al. 2004), who found elasticity of the car stock with respect to fuel price ranging from -0.3 (LT) to -0.1(ST) and with respect to capital cost ranging from -0.49 (LT) to -0.24 (ST).

$$\frac{CAP_t}{CAP_{t-1}} = (\frac{FP_t}{FP_{t-1}})^{\alpha_{FP}} \cdot (\frac{CC_t}{CC_{t-1}})^{\alpha_{IC}} \cdot (\frac{GDP_t}{GDP_{t-1}})^{\alpha_{y\_t}} \quad (6\text{-}11)$$

$$\alpha_{FP} = -0.2 \quad \alpha_{IC} = -0.5 \quad \alpha_{y\_2010} = 1$$

$CAP_{(t)}$......number of cars in the fleet
CC ... capital costs (fixed costs)
FP ... fuel price
GDP ... gross domestic product
α_{FP} ... fuel price elasticity (fleet)
α_{IC} ... elasticity on fixed costs (fleet)
α_y ... income elasticity (fleet)

6.4.2 Modeling of the car user behavior

Shifts in economic framework conditions have short run influence on the behaviour of consumers. Car owners react to changes in the cost of energy service by adapting their use intensity expressed in kilometres travelled per year. This correlation is modelled by elasticities of fuel price and income. For the calibration of the model statistic data (1990-2008) was used. The derived elasticity of driving distance with respect to income was 0.3 and -0.3 with respect to fuel price. This is within the range of results found by (Johansson & Shipper 1997), where elasticity of mean driving distance with respect to fuel price ranges from -0.35 to -0.05 and with respect to income from -0.1 to 0.35. The more recent analysis of (Goodwin et al. 2004) underlines the validity of the elasticity with respect to fuel price (-0.3 for LT).

$$\frac{D_t}{D_{t-1}} = (\frac{FP_t}{FP_{t-1}})^{\omega_{FP}} \cdot (\frac{GDP_t}{GDP_{t-1}})^{\omega_Y} \tag{6-12}$$

$\omega_{FP} = -0.3 \qquad \omega_Y = 0.3$

$D_{(t)}$ ... distance travelled by year
FP ... fuel price
GDP ... gross domestic product
ω_{FC} ... price elasticity (driving distance)
ω_Y ... income elasticity (driving distance)

6.4.3 Transport service level – shares of vehicle classes

The specific service costs not only affect the annual car sales but also the service level of the cars sold. In the model the service level is defined by the average characteristics of cars (average vehicle weight and engine power) (see. chapter 3.6).

The model defines three categories of vehicles, compact class, middle class, upper class, which are also the options consumers are choosing from when purchasing a car (see Table 6-1). The specifications are set in such way as to represent the Austrian passenger car fleet. Each class is defined by mass and engine power and a minimal required driving range. The current customer preferences concerning vehicle categories in Austria were determined by historical data sets derived from statistical data (Statistics Austria 2009b) and are used as a basis for the model settings.

Table 6-1: Passenger car classes and user categories

vehicle classes:	reference weight [kg]	reference power [kW]	user groups:	kilometrage [km year-1]
compact class	1.000	50	weekend user	10.000
middle class	1.500	75	regular user	15.000
upper class	2.000	120	commuter	20.000

The effect of fuel price and income on the mean specifications of cars is modelled by introducing a service factor F_t representing average mass and engine power of the cars sold. It is assumed that the specifications of sold cars are distributed around the mean value F_t in a distribution with positive skew (see Figure 6-4). The distribution was defined based on statistic data on passenger car sales in Austrian (see Appendix B). In the model the distribution is used to derive the shares of the three vehicle classes (see equation (6-14)). The development of this parameter F_t is determined by Income and fuel prices (see equation (6-15)).

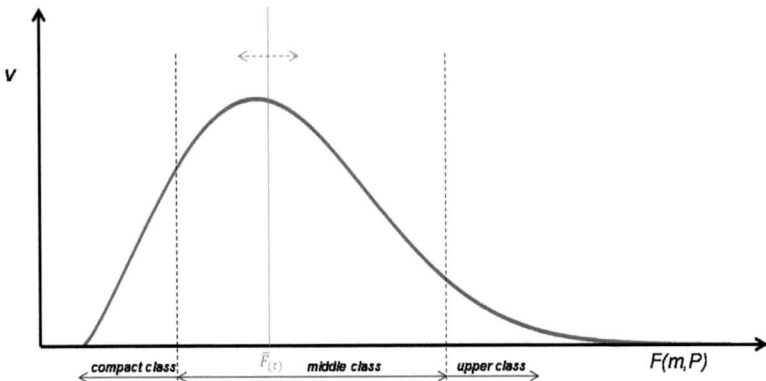

Figure 6-4: Distribution of the service factor within car sales (schematic illustration)

$$\overline{F_t} = f(\overline{m_t}, \overline{P_t})$$ (6-13)

$$z_{i(t)} = Z_t^{-1} \cdot \int_{F_{i\_l}}^{F_{i\_u}} v(F; \overline{F}_{(t)}) dF$$ (6-14)

$\overline{F}_{(t)}$ ... mean vehicle service factor
$\overline{m}$ ... mean vehicle mass
$\overline{P}$ ... mean vehicle power
$z_{i(t)}$ ... share of the vehicle class i
$Z_{(t)}$ ... new cars registered per year
F_i ... specification of the vehicle class
F_{i_u} ... maximum service level of vehicle class i
F_{i_l} ... minimum service level of vehicle class i
v ... distribution of sold vehicles around the mean value (=vehicle service factor)

$$\frac{F_{(t)}}{F_{(t-1)}} = (\frac{FP_{(t)}}{FP_{(t-1)}})^{\beta_{FP}} \cdot (\frac{CC_{t}}{CC_{t-1}})^{\beta_{IC}} \cdot (\frac{GDP_{(t)}}{GDP_{(t-1)}})^{\beta_{Y}}$$

(6-15)

$\beta_{FP} = -0.3 \quad \beta_{Y} = 0.3 \quad \beta_{IC} = -0.5$

β_{FP} ... fuel price elasticity
β_{IC} ... elasticity with respect to fixed costs
β_{Y} ... income elasticity
FP ... fuel price
GDP ... income

The corresponding elasticities were determined through calibration runs comparing the model-results with historic data on Austrian passenger car sales in the time frame 1990-2008 (see Figure B-3 and Figure B-4 in Appendix B). Also the distribution v representing the allocation of vehicle specification around the annual mean value F_t was determined based on statistic data (data sources: (Salchenegger 2006) (Pötscher 2009)). In the last decades there has been a continuous increase in both weight and power of passenger cars in Austria (Meyer & Wessely 2009) which points out the influence of income on the characteristics. The higher prices between 2003 and 2009 lead to a saturation of the average weight which is an indicator of considerable price sensitivity of this parameter.

There is hardly any historic top-down analysis covering the effects of price and income on vehicle characteristics. Most studies only capture the fuel intensity of the cars in the fleet (e.g. (Johansson & Shipper 1997)). Even though there is certainly a correlation between vehicle size or engine power and fuel consumption, the parameter is not applicable in the model since it also implies technological improvements of cars that are captured separately in this model.

6.5 Bottom-Up Fleet Model

The bottom-up model of the passenger car fleet represents a central element of the whole simulation model. In the model the fleet is divided into three vehicle classes. Within these classes there are different user groups each with a specific yearly kilometrage (see Table 6-1). There are also different vehicle propulsion systems in the stock and vehicles have dissimilar levels of average efficiency depending on their year of construction. Moreover, there is a detailed coverage of vehicle efficiency and technologies in the fleet model. To capture all these characteristics in the model detailed historic data on the Austrian passenger car market was used (data source: (Statistics Austria 2009b)).

6.5.1 Modelling the exchange rate of cars

The passenger car fleet reacts very slowly to shifts in framework conditions. This is due to its inherent low exchange rate of cars. Once registered a car statistically remains in the fleet between 10 to 15 years. To represent this inertness correctly in the model a stochastic approach is applied.

The actual fleet CAP_t is determined by the surviving cars of all previous cohorts of car. In the model this implies 30 cohorts of cars. The fleet structure can be expressed as follows:

$$CAP_t = \sum_n SZ_n \qquad (6\text{-}16)$$

CAP ... car fleet
SZ_n ... survivors of a vehicle cohort n.

Every year a certain amount of cars of the stock is being decommissioned. The most common reason for setting a car out of service is the occurrence of mechanical failure. When the repair of the failure cannot be justified under the given economic framework conditions in the specific country or region the car is either scrapped or sold abroad. Yearly decommission rates are modelled using the weibull distribution. This distribution is commonly used to determine the likelihood of mechanical failure and is also used in comparable models to represent the annual scrapping of cars in the fleet (Zachariadis 2005) (Christidis et al. 2003) (MIT 2008).

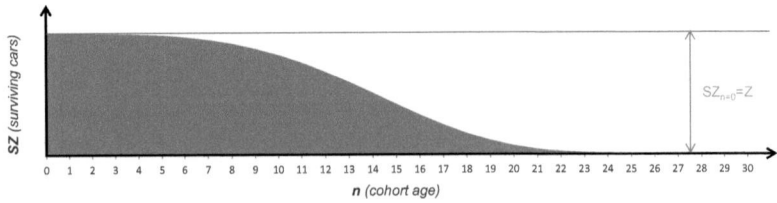

Figure 6-5: Surviving cars in the cohorts

$$SZ_{n(t)} = SR_n \cdot Z_{t-n} \tag{6-17}$$

$$SR(n,s,\lambda) = 1 - e^{-(x/\lambda)s} \tag{6-18}$$

SZ_n ... survivors of a car generation
SR ... survival rate
Z ... number of cars newly registered
n ... index of car cohort
s, λ ... parameters of the weibull distribution

The survivors SZ of a car cohort with the age n are determined by the survival rate SR for cars with the age n. The Survival Rate SR is modelled through the cumulative distribution function of the weibull distribution.
The number of cars registered every year depends on the development of the overall fleet and the number of cars scrapped. The fleet development is given by the top down demand described in chapter 6.4.1.

$$Z_t = CAP_t - CAP_{t-1} + Z_{SCRAP(t)} \tag{6-19}$$

$$Z_{SCRAP(t)} = CAP_{t-1} - \Sigma SZ_t \tag{6-20}$$

Z_t ... new cars registered per year
$Z_{SCRAP(t)}$ ... cars scrapped per year

The variables k and λ are determining parameters of the Weibull function and were determined using statistic data of the Austrian passenger car fleet. Thereby, the modelled fleet CAP_t has to correspond with the real vehicle stock CAP_real_t recorded by the federal statistic organisation and secondly the modelled average age of the vehicle fleet has to correspond with the average age of the fleet published by statistic organisation. The modelled fleet has to fulfil the following conditions:

$$\forall \quad k, \lambda = const : \quad CAP_t = CAP\_real_t \wedge \overline{a}_t = \overline{a}\_real_t \tag{6-21}$$

CAP_real ... car fleet according to statistics
$â$ ... average car age in the fleet
k ... parameter of the Weibull distribution.
λ ... parameter of the Weibull distribution.

Also the user groups with their different yearly kilometrage and their corresponding fuel consumption were determined based on statistics on transport fuel consumption in Austria (data source: (Fachverband Mineralölindustrie 2010a)). The parametrization is set in such a way that the modelled fuel consumption of the fleet corresponds with real one reported by Austria statistics:

In the case of Austria this way of calibration implicates some uncertainties as gasoline and especially diesel fuels are not only used by passenger cars but also by light duty vehicles and trucks. Furthermore, considerable amounts of both gasoline and diesel are consumed by foreign vehicles and not by the Austrian fleet. To determine the quantities correctly additional data from the Austrian federal statistics on household energy consumption (Statistics Austria 2009a) and the Austrian federal ministry of environment have been included (Schneider & Wappel 2009).

$$EC\_CAP = EC\_CAP\_real \qquad (6\text{-}22)$$

$$EC\_CAP_h = EC\_CAP\_real_h \qquad (6\text{-}23)$$

p_u [%] p_{ju} [%]

EC_CAP ... cumulative energy consumption in the fleet
p_u ... share of user groups
p_{ju} ... share of technologies in user groups

6.6 Energy Consumption and greenhouse gas emissions of the passenger car fleet

From the detailed fleet model the cumulative energy consumption and the greenhouse (GHG) gas emissions can be derived. The model distinguishes between well-to-wheel (WTW) and tank-to-wheel (TTW) emissions. For passenger cars, energy consumption and GHG-emissions are often given in the TTW view in litres per 100 km and g CO_2 per km. In this analysis both the TTW and the WTW balances for energy consumption and greenhouse gas emissions are calculated. The TTW energy consumption is calculated as follows:

$$EC\_CAP_t = \sum_{i,j,n,u} FC_{i,j,n} \cdot CAP_{i,j,n,u} \cdot D_{i,j,n,u} \qquad (6\text{-}24)$$

EC_CAP ... cumulative energy consumption in the fleet
FC ... specific fuel consumption of cars
D ... yearly driving distance

The final energy consumption of the entire fleet is determined by the specific consumption of the vehicles *FC* (of different classes *i*, with different technologies *j* and different vintages *n*) the vehicle stock *CAP* and the kilometres travelled per year *D* (in different user categories *u*).

Analogically the cumulated GHG emissions *CUM_GHG* are modelled:

$$GHG\_CAP_t = \sum_{ijnu} GHG_{i,j,n} \cdot CAP_{i,j,n,u} \cdot D_{i,j,n,u} \tag{6-25}$$

GHG_CAP ... cumulative greenhouse gas emissions in the fleet

For an unbiased view on energy consumption and greenhouse gas emissions the entire energy conversion pathway represented in the well-to-wheel balances has to be considered (see chapter 8.4). The corresponding cumulative energy consumption *CUMEC_WTW* and GHG emission *CUMGHG_WTW* are calculated as follows:

$$EC\_WTW\_CAP_t = EC\_CAP + \sum_h EC\_CAP_h \cdot CE_h + \sum_{i,j} Z_{ij} \cdot CE_{ij} \tag{6-26}$$

$$GHG\_WTW\_CAP_t = GHG\_CAP + \sum_h GHG\_CAP_h \cdot GE_h + \sum_{i,j} Z_{ij} \cdot GE_{ij} \tag{6-27}$$

EC_WTW_CAP ... cumulative well-to-wheel energy consumption of the fleet
CE ... well-to-wheel energy per unit
GHG_WTW_CAP ... cumulative well-to-wheel greenhouse gas emissions of the fleet
GE ... well-to-wheel greenhouse gas emissions per unit

7 Assumptions for Scenario Development

The model described in the previous chapter was used to analyse the effects of fuel price increases, technological progress in propulsion systems and policies on the passenger car fleet in the time frame 2010-2050. To evaluate the effect of these changes scenarios are developed based on different assumptions concerning fossil fuel price development and degree of political intervention.

The focus of the scenarios is set on electrification of passenger cars including the entire range from conventional to fully electric cars. Even though hydrogen fuel cell technology is foreseen as a technology-option in the model it is not considered in these particular scenarios. As described in chapter 5 the future cost development of fuel cell cars is uncertain. Given the serious technical, economical and infrastructural barriers the technology has to face it will rather become a long term option for passenger car transport beyond the time horizon of the analyzed scenarios.

7.1 Global fossil fuel price levels

In the next decades economic growth and mass motorisation of developing countries (e.g. China, India) will lead to a further growth in crude oil demand while production of conventional oil is decreasing (IEA 2009b). This development will lead to a steady increase in fossil fuel prices.

For the analysis two fuel price scenarios are used, that show different price increases in the time frame 2010-2050:

Low-Price-Scenario: in this scenario it is assumed that the fossil fuel price increases moderately leading to a doubling of crude oil price up to 2050. This assumption is based on the "PRIMES-high" energy price scenario (Kapros et al. 2008).

High-Price-Scenario: in this scenario it is assumed that crude oil price increases much stronger leading to a tripling of the price up to 2050. This would mean an oil price of approximately 225 $ bbl^{-1}. (see Figure 7-1)

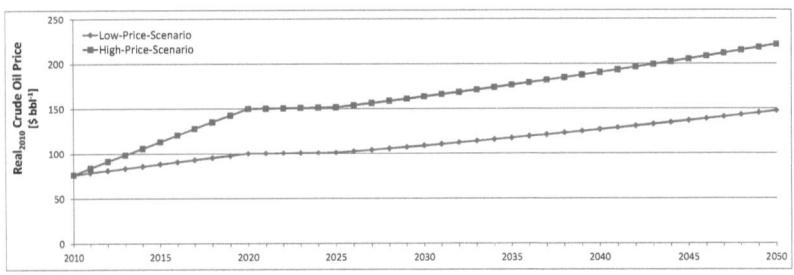

Figure 7-1: Real$_{2010}$ Oil Price Scenarios

Figure 7-2 and Figure 7-3 show the real net price developments of energy carriers for passenger car transport in the two price scenarios. In the case of gasoline, diesel and CNG the real price implies the net product price and the margin of the petrol station and does not include fuel taxes and value added tax. In the case of electricity this implies the net energy price and grid tariffs as given by the Austrian regulator (e-control 2010b). The derived gross electricity price in 2010 is 17 €cent kWh^{-1}.

In the scenarios the oil based products are assumed to be affected by the crude oil price increase stronger than electricity. Natural gas based CNG is assumed to show a similar net price development as gasoline and diesel, as its price is traditionally linked to the crude oil price. Electricity is much more diversified which makes it less sensitive to oil price shocks.

There is some uncertainty in electricity prices for electric cars since charging infrastructure has to be considered as well. In the early stages of electric vehicle diffusion, infrastructure cost can be neglected since early adopters will be consumers who have a parking space with plug available. However, when it comes to large scale diffusion a public charging infrastructure will have to be built. The cost of such an infrastructure will be included in the electricity price for electric cars. (Fraunhofer ISI 2010) analysed specific infrastructure cost for electric vehicle charging finding that specific costs of public infrastructure would range from 2 to 3 €cent kWh^{-1} if 3.5 kW charging is applied. In the case of home charging infrastructure cost would even be lower (1 €cent kWh^{-1}). The electricity price assumed in the analysis also implies cost of charging infrastructure.

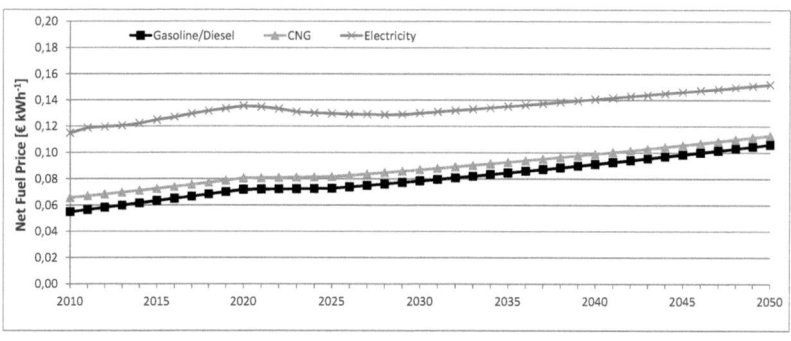

Figure 7-2: Real$_{2010}$ Net Fuel Prices in the "Low-Price-Scenario"

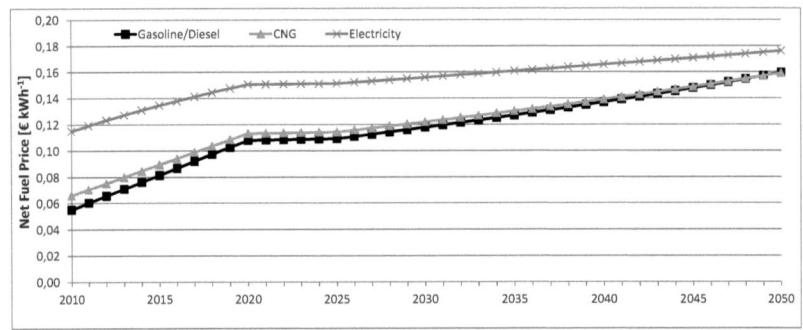

Figure 7-3: Real$_{2010}$ Net Fuel Prices in the "High-Price-Scenario"

7.2 Political Framework Conditions in Austria 2010-2050

The intended GHG reductions in Austrian road transport can only be achieved by a combination of more efficient propulsion technologies, low carbon fuels and generally a more efficient use of vehicles. The achievement of these goals will require enforced policy intervention in both passenger and freight transport in the next decades.

As demonstrated in chapter 5.8 taxes have considerable impact on the cost of passenger car mobility in Austria. By affecting the costs of transport, policies also affect the entire structure of the passenger car sector and the overall demand for this transport mode (cf. chapter 3). In addition, they can influence the market share of different vehicle propulsion technologies and thereby have an effect on energy consumption and greenhouse gas emissions of the sector.

The effects of policy strategies on the passenger car fleet in Austria are analysed through scenarios. In these scenarios the effectiveness of different policy measures is analysed in an environment of increasing fossil fuel prices. The results should demonstrate how political and economic framework conditions affect the development of the passenger car fleet in terms of energy consumption, energy carriers, efficiencies and greenhouse gas emissions up to 2050.

The analysis includes two main policy scenarios: One with no additional policy measures compared to 2010 and one with an active policy strategy with the objective to improve efficiency and cut GHG emissions. Furthermore, two sub-scenarios are developed analysing the effects of fuel specific and vehicle specific taxation measures separately.

The measures in both scenarios are implemented between 2010 and 2020 and should thereby demonstrate the importance of policy measures within the next decade for the mid- to long-term development of the passenger car fleet and its corresponding energy consumption and GHG emissions.

7.2.1 Business as usual (BAU) Scenario

In this scenario political framework conditions remain comparatively the same to the status of 2010. The only change is a slight adjustment of the fuel tax, taking into account that CNG would be taxed with the same rate as diesel starting in 2015 (see table Table 7-1 and Table 7-2). The BAU scenario gives an outlook on the development of the Austrian passenger car fleet if no additional policy measures such as fuel tax increases or vehicle taxation are taken in the upcoming decade.

7.2.2 Policy scenario

The Policy Scenario should demonstrate how political framework can help reduce GHG emissions of the passenger car fleet in order to contribute to the country's emission reduction commitments.

In this scenario major changes in political framework conditions are adopted between 2010 and 2020. Taxes are adapted with a clear focus on increasing energy efficiency and reducing GHG emissions. The instruments that are used are fuel taxes and tax on acquisition of cars.

Taxes on gasoline and diesel are being raised stepwise between 2010 and 2020 and tax on acquisition is being adapted in order to promote sales of efficient cars. One possible way to implement that measure in the Austrian tax on acquisition scheme (see chapter 2.6.1.1) would be by lowering the upper threshold for greenhouse gas emissions from 160g km^{-1} (status 2010) to 100g km^{-1} in 2016. (see Table 7-1 and Table 7-2). This measure would especially affect cars and technologies with higher greenhouse gas emissions making them significantly more costly.

As described in chapter 2.6.1 electric cars are excluded from tax on acquisition in Austria. Since the existing scheme is calculated on the basis of the gasoline or diesel consumption and the TTW greenhouse gas emissions, taxation of electric cars would require new mechanisms. In the Policy scenario the implementation of the 100g km^{-1} threshold in 2016 also implies the introduction of a 2% tax on acquisition for electric cars.

Table 7-1: Political framework conditions within the two scenarios

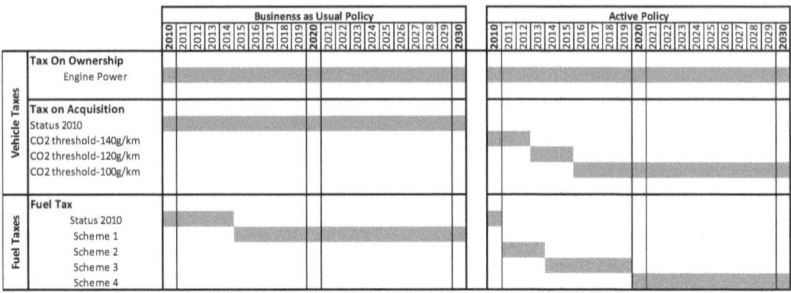

Table 7-2: Fuel taxation schemes

		Status 2010	Scheme 1	Scheme 2	Scheme 3	Scheme 4
Gasoline	€ kWh-1	0,051	0,051	0,05	0,07	0,10
Diesel	€ kWh-1	0,036	0,036	0,05	0,07	0,10
CNG	€ kWh-1	0	0,036	0,05	0,07	0,10
Electricity	€ kWh-1	0	0	0	0	0,02

The two fuel price scenarios and the two policy scenarios add up to the following four main scenarios:

- "Business as usual" Policy & moderate fossil fuel price increase (*BAU & Low Price-Scenario*)

- "Active" Policy & moderate fossil fuel price increase (*Policy & Low Price-Scenario*)

- "Business as usual" Policy & substantial fossil fuel price increase (*BAU & High Price-Scenario*)

- "Active" Policy & substantial fossil fuel price increase (*Policy & High Price-Scenario*)

7.3 Specific Service Costs 2010-2050

Economic competitiveness is one of the crucial factors for the future development of hybrid and electric vehicles. In this analysis specific service cost are considered as the most important decision criteria for the choice of a vehicle powertrain systems. A detailed overview on the economic performance of the different propulsion technologies without considering political framework conditions was already given in chapter 5.5. In this chapter the influence of the policy scenarios defined above on the cost of the different propulsion technologies will be illustrated.

As described in chapter 6 the specific service cost are the central parameter for the development of the passenger car sector in terms of fleet size, propulsion technologies, vehicle characteristics and intensity of vehicle use.

In the model the specific costs of all powertrain options within the three vehicle classes and the different user groups are calculated dynamically for the time frame 2010-2050, considering shifts in fuel prices, technological costs, taxation and income.

The specific service costs SC of each vehicle of the vehicle class i, with the technology j are determined by their specific fuel costs FC, specific fixed operations costs OC and specific capital costs CC as follows (cf. chapter 3.8.2):

$$SC_{ij} = CC_{ij} + OC_{ij} + FC_{ijh} \ [€\ km^{-1}] \qquad (7\text{-}1)$$

To calculate the specific service costs a standard depreciation time of 10 years and an interest rate of 5% are used. It is evident that the economic performance of a propulsion system depends on the yearly driving distance of the user and differs among the defined user categories (see Table 6-1 in chapter 6.4).

Figure 7-4 and Figure 7-5 illustrates the gasoline and electricity price increase corresponding to the four main scenarios, resulting from the combination of the two fossil fuel price scenarios and the two policy schemes. Especially the *Policy* scenarios lead to a substantial increase in gasoline and diesel prices that cause a doubling of the prices in 2020 compared to 2010. Even without higher taxes the global fossil fuel price increase leads to a 50% increase of gasoline and diesel prices up to 2020.

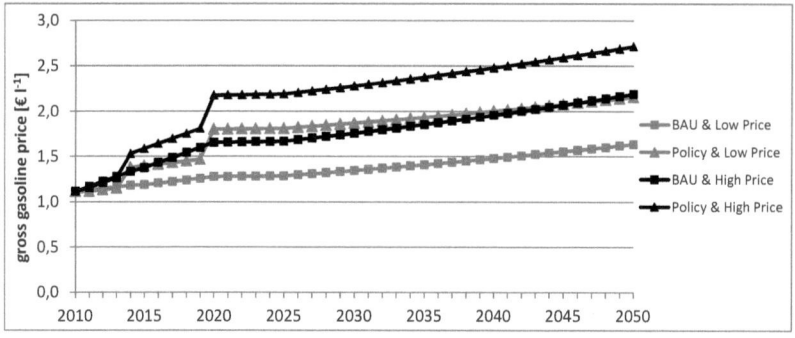

Figure 7-4: Development of real$_{2010}$ gross gasoline prices in the four main scenarios

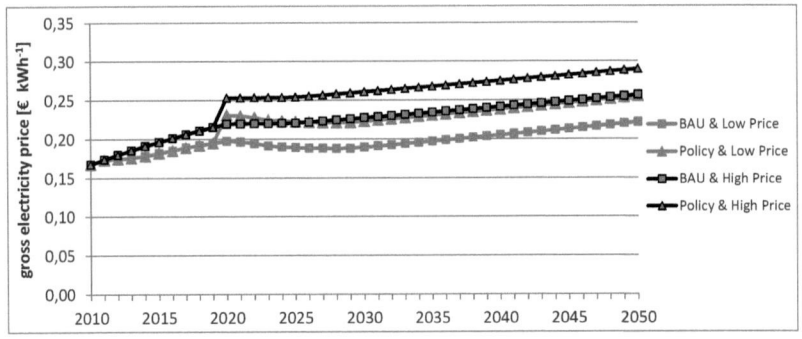

Figure 7-5: Development of real$_{2010}$ gross electricity prices in the four main scenarios

The different schemes of political intervention in the two scenarios are strongly reflected in the specific service cost of the propulsion technologies in the time frame 2010-2050. Figure 7-6 and Figure 7-7 show the corresponding development in the *"BAU"* and in the *"Policy"* scenario for compact class cars with a yearly driving distance of 10 000 km. The charts illustrate the impact of the policy measures on the cost of all gasoline and diesel based propulsion systems. Taking a look at the *BAU scenario* results it is noticeable that the assumed price increase gets almost completely compensated by the efficiency improvement of cars, which means that the real cost of transport remains quite constant. Due to the cost reductions in batteries and electric drive components the EV becomes the least cost option after 2030 in the *BAU scenario*. In the Policy scenario the EV becomes the least cost option after 2020 with significantly lower cost the following years and decades. The long term cost development of the EV might necessitate further policy measures after 2020 to raise their real cost in order to avoid rebound effects.

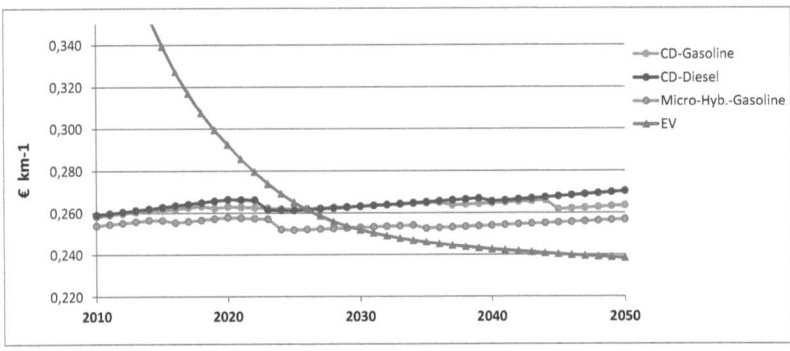

Figure 7-6: Specific service cost of compact class cars: *BAU & Low Price – Scenario* (10 000 km year $^{-1}$)

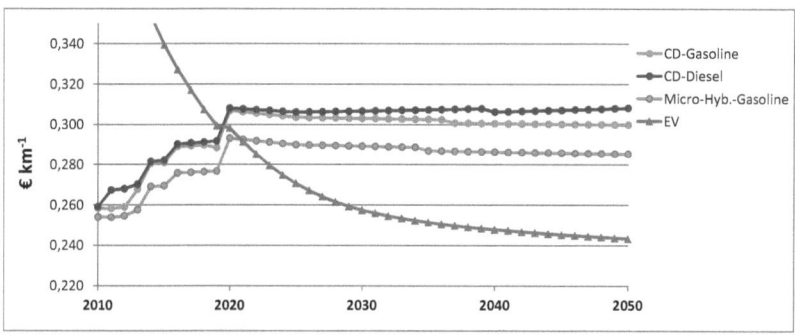

Figure 7-7: Specific service cost of compact class cars: *Policy & Low Price – Scenario (10 000 km year^{-1})*

Figure 7-8 and Figure 7-9 show the cost of propulsion systems for both scenarios in the middle class segment at a yearly driving distance of 15 000 km. In the BAU scenario the development shows the same characteristic like in the compact class: the price increase of fossil fuels is to a large extent compensated by the improved efficiency of the propulsion systems. Even though the cost of EVs and PHEVs decrease considerably, mild hybrids remain the least cost option until 2035.

In the Policy scenario the increase in fuel and vehicle taxes leads to considerable increase in cost of ICE based propulsion technologies and thereby to a short- to mid-term increase of transport service cost. In this scenario PHEVs and BEVs become the least cost option at around 2020. Just like in the compact class the long term cost decrease of electric propulsion technologies might necessitate further policy measures to prevent rebound effects.

It is evident that the assumed yearly driving distance affects the overall cost of all cars as well as the cost ranking of propulsion technologies. Longer yearly driving distances favour more efficient and therefore more costly technologies. Consequently, the three user categories in the model with 10 000, 15 000 and 20 000 km of yearly driving distance show different cost rankings (see Figure B-9 to Figure B-12 in Appendix B).

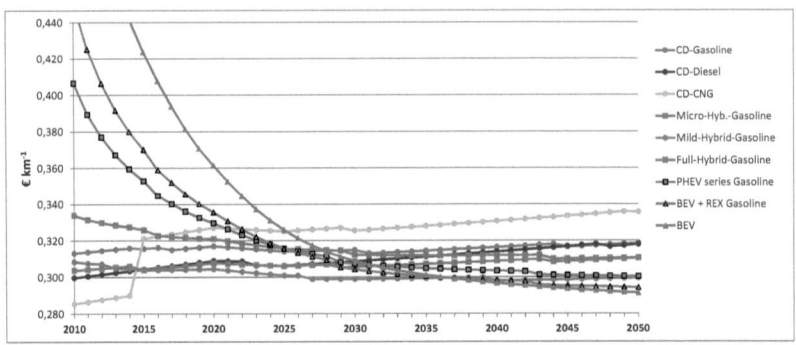

Figure 7-8: Specific service cost of middle class cars: *BAU & Low Price – Scenario* (15 000 km year^{-1})

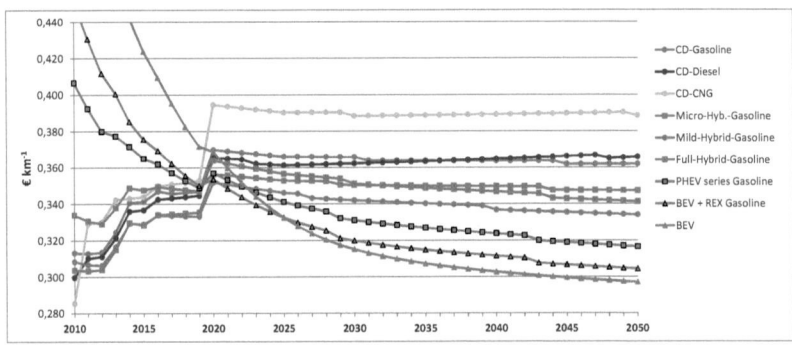

Figure 7-9: Specific service cost of middle class cars: *Policy & Low Price – Scenario* (15 000 km year^{-1})

Figure 7-10 shows the comparison of yearly cost split up in cost categories in the middle class in 2030 in the *BAU* and in the *Policy scenario*. It illustrates the significant impact of fuel tax and tax on acquisition on the overall cost. In the Policy scenario the increased taxes lead to a clear advantage in costs for PHEVs and BEVs. The trend toward diesel cars in the 1990ies showed that even these small differences could have an impact on the market shares of technologies (cf. chapter 2.4).

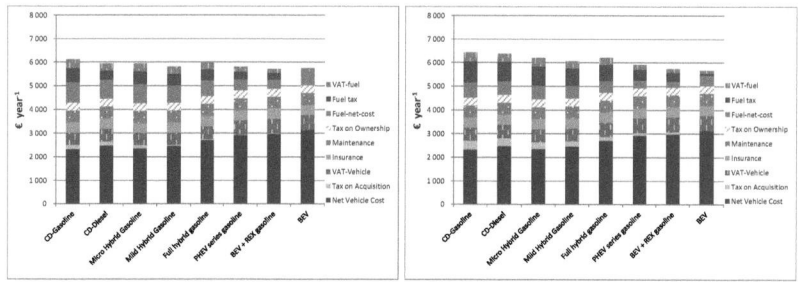

Figure 7-10: Total yearly cost of middle class cars with different propulsion systems in 2030: *BAU & Low Price* and *Policy & Low Price-Scenario* (15 000 km year^{-1})

Figure 7-11 and Figure 7-12 show the development of specific service cost for middle class cars for the *BAU* and the *Policy* scenario with strong fossil fuel price increases (*High Price*). Even without higher taxes the specific cost of transport increases considerably. This development promotes efficient and electric propulsion technologies.

In the corresponding *Policy* scenario the combination of increasing fossil fuel prices and higher taxes on vehicles and fuels leads to an increase in specific transport cost of conventional cars by almost 25 % up to 2020. Under this framework conditions PHEV and BEV are by far the least cost technologies after 2020.

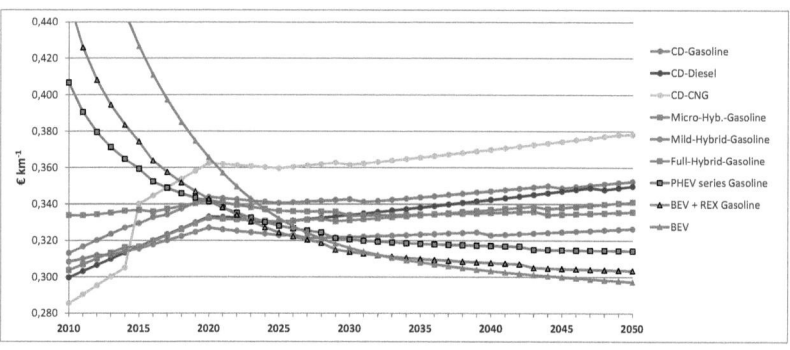

Figure 7-11: Specific service cost of middle class cars: *BAU & High Price – Scenario* (15 000 km year^{-1})

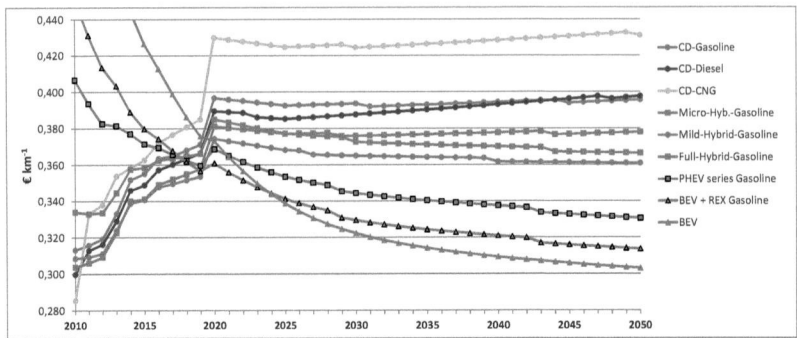

Figure 7-12: Specific service cost of middle class cars: *Policy & High Price – Scenario* (15 000 km year $^{-1}$)

7.4 Market- and Fleet-Penetration 2010-2050

In the following chapters the market- and penetration of the four main scenarios will be illustrated:

- "Business as usual"-Policy & moderate fossil fuel price increase (*BAU & Low Price-Scenario*)
- "Active" Policy & moderate fossil fuel price increase (*Policy & Low Price-Scenario*)
- "Business as usual"-Policy & substantial fossil fuel price increase (*BAU & High Price-Scenario*)
- "Active" Policy & substantial fossil fuel price increase (*Policy & High Price-Scenario*)

7.4.1 Business as usual policy & low fuel price scenario (BAU & Low Price)

In this scenario no considerable measures to promote efficient and alternative vehicle technologies are taken. The development of market shares of propulsion technologies illustrated in Figure 7-13 shows a strong trend towards hybrid cars. This development is mainly driven by the improving economic competitiveness of hybrid powertrain systems in an environment of increasing fuel prices and by cost reduction resulting from learning effects of key components. This leads to a substitution of conventional powertrain systems by micro and mild hybrid systems. Both technologies are closely related to conventional powertrain systems and can increase vehicle efficiency at relatively low additional cost.

In a mid to long term period technologies with a higher degree of electrification, e.g. PHEVs and BEVs, can only gain market shares in a slow pace. Figure 7-14 shows the corresponding development of the vehicle fleet in the *BAU-Scenario*.

The vehicle fleet is growing constantly in the time frame 2010 – 2050 (+27 % up to 2030; +47 % up to 2050). The demand for transport and the service level keeps increasing as a result of its relatively low cost, which is reflected in the growth of the car fleet. The increasing crude oil price that is assumed in this scenario is compensated by the improved efficiency of cars, keeping overall price of transport low. In 2030 PHEVs and EVs together account for 12 % of the car fleet growing to 55 % up to 2050.

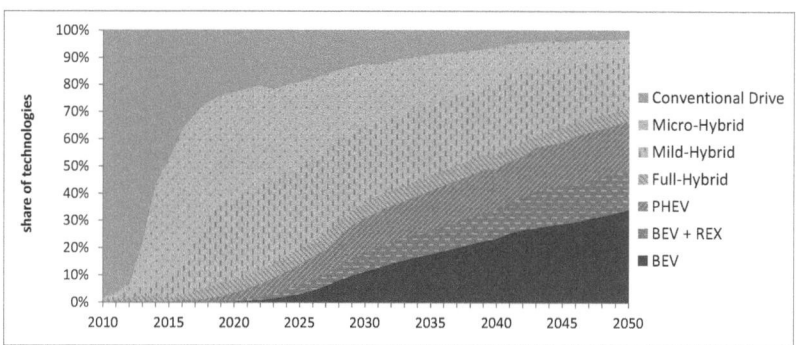

Figure 7-13: Market shares of propulsion technologies within yearly car sales: *BAU & Low Price – Scenario*

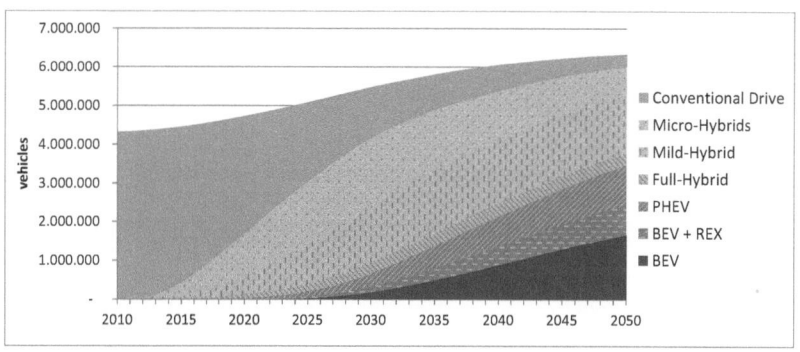

Figure 7-14: Passenger car fleet: *BAU & Low Price – Scenario*

7.4.2 Active policy & low fuel price scenario (Policy & Low Price)

In the *Policy Scenario* higher taxes on fuels combined with tax reduction for efficient vehicles lead to an improved competitiveness of electric propulsion technologies (see Figure 7-10).

In a short term period there is a similar development of hybridisation as in the *BAU scenario* with micro and mild hybrids massively gaining market shares. Starting 2020, there is a significant trend toward electric powertrain systems leading to a market share of electric cars of 68 % in 2030. Plug-In Hybrid (PHEVs) and extended range electric vehicle (BEV+REX) account for another 25 % of the market (see Figure 7-15). Another reason why these high market shares can be achieved in this scenario is the fact that the assumed framework conditions also lead to a change in the average use of passenger cars which also affects the consumers' expectations concerning necessary electric driving range. Also the technical characteristics of batteries are expected to improve making electric cars more attractive.

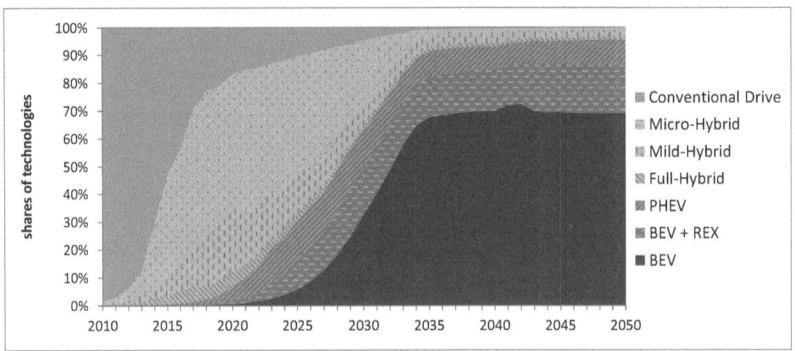

Figure 7-15: Market shares of propulsion technologies within yearly car sales: *Policy & Low Price – Scenario*

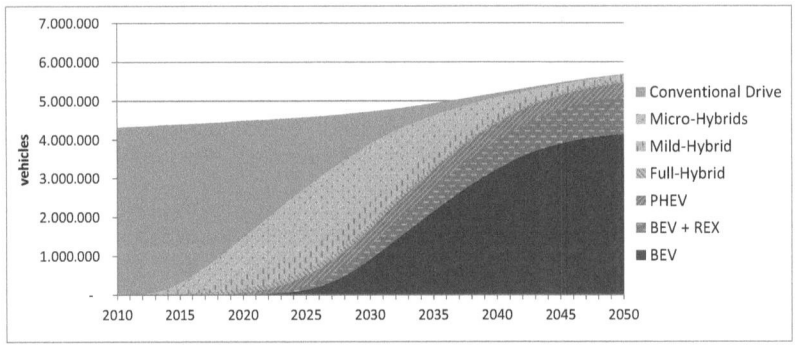

Figure 7-16: Passenger car fleet: *Policy & Low Price – Scenario*

In general, taxation of cars and fuels drives overall transportation costs, which reduces the demand for passenger transport. This development causes a deceleration of vehicle fleet growth (*up to 2030:* +9 %; *up to 2050:* +31 %). The pick-up of growth in the last two decades is and indicator for the rebound-effect that is caused by the relatively low cost of electric propulsion systems (cf. chapter 3.7 & 7.3).

Conventional drive systems are being replaced by hybrid systems in a short- to mid-term period. In a long-term period electrified cars like PHEVs and BEVs gain a considerable share in the overall vehicle fleet. Together they reach a share of more than 90 % of the vehicles fleet in 2050 (*2030:* 36 %; *2050:* 96 %) (Figure 7-16).

7.4.3 Business as usual policy & high fossil fuel price scenario (BAU & High Price)

This scenario combines the passive policy of the *BAU scenario* with a stronger increase of the fossil fuel price level. In this scenario the strong increase of fuel prices cannot be outweighed by the improvement of vehicle efficiency of the propulsion technologies (see Figure 7-17 in chapter 7.3). The resultant higher transport costs in this scenario lead to a deceleration of overall transport demand growth and a lower service level. This causes a lower fleet growth (*up to 2030:* +16 %; *up to 2050:* +33 %), less intense use of cars and smaller/less powerful cars sold. The higher cost also drives more consumers to adopt alternative vehicle propulsion technologies, leading to stronger diffusion of electric powertrain systems than in the low price case (*2030:* 24 %; *2050:* 88 %).

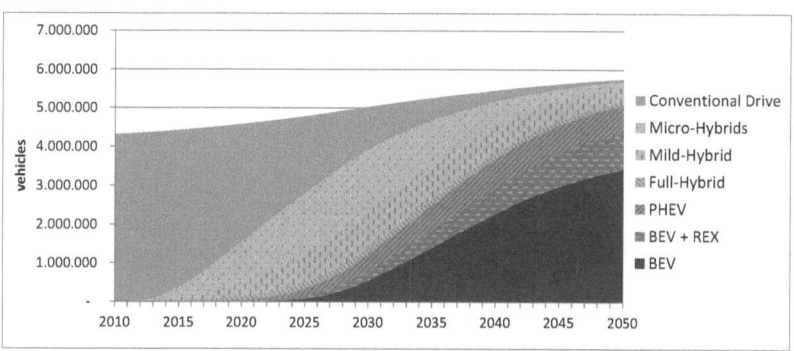

Figure 7-17: Passenger car fleet: *BAU & High Price – Scenario*

7.4.4 Active policy & high fossil fuel price scenario (Policy & High Price)

As indicated by the cost development the given framework conditions in this scenario strongly promote efficient vehicle propulsion technologies and slows down the growth in transport demand. This is strongly reflected in the car fleet development. In this scenario the fleet almost stabilizes at the 2010 level. Only in a long term there is a further growth driven by the decreasing cost of electric propulsion systems and the resulting rebound effect (*up to 2030:* +3 %; *up to*

2050: +26 %). The share of PHEV and EVs reaches 42 % in 2030 and 99 % in 2050.

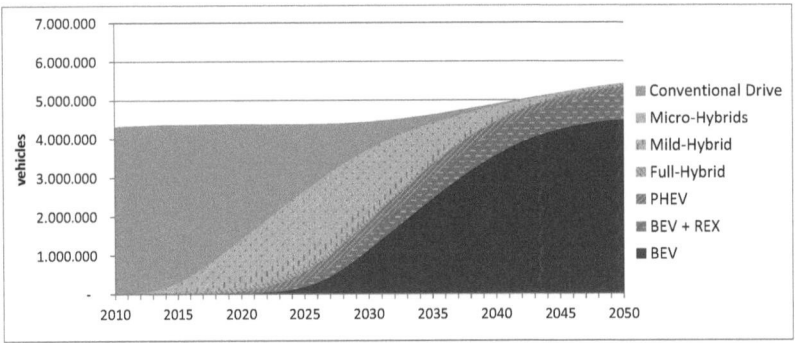

Figure 7-18: Passenger car fleet: *Policy & High Price – Scenario*

7.4.5 Impact of technological learning effects

It is evident that cost reduction of batteries has a strong impact on the cost effectiveness of hybrid and electric powertrain systems and thereby on their future market shares. The cost reductions are modelled with the help of technological learning effects and therefore depend on the used learning parameter (see chapter 5.6). A sensitivity analysis with respect to the learning rate is performed to analyse the resulting uncertainty. The results for the *BAU Scenario* show that the sensitivity of the technology market shares to variations in the learning rate is critical in the mid- to long-term, especially with respect to the diffusion of all electric cars (see Figure 7-19). In the *Policy Scenario* the impact of the learning rate is less significant (Figure 7-20). In order to take this uncertainty into account effects of the learning rate is depicted in all energy- and greenhouse-gas-related scenarios in chapter 8.

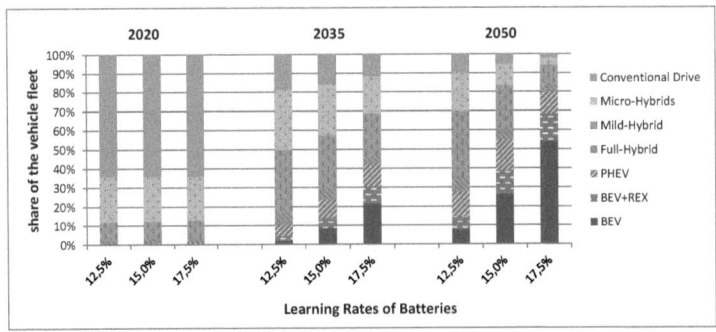

Figure 7-19: Sensitivity of technology diffusion with respect to learning effect – *BAU-Scenario*

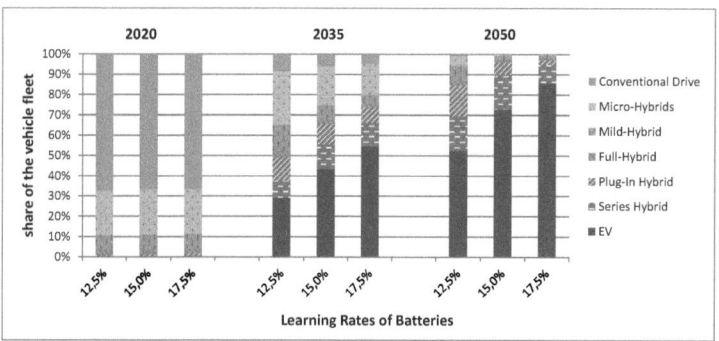

Figure 7-20: Sensitivity of technology diffusion in yearly car sales with respect to learning effect – *Policy-Scenario*

7.4.6 Average characteristics of passenger car sales

The effects of political and economic framework conditions also affect the transport service level which is reflected in the average characteristics of cars sold (cf. chapter 3.6 & 6.4.3). Figure 7-21 to Figure 7-23 illustrate the effect of the analysed policy and fuel price scenarios on the characteristics of new cars in Austria 2010 - 2030. It shows how political framework conditions affect the average service level of cars reflected in their mass and engine power which also affects their average fuel consumption and GHG emissions.

In the *BAU Scenario* with low fossil fuel price increase the average power remains relatively the same and vehicle mass slightly decreases as a consequence of enhanced use of light-weight materials.

In the *Policy Scenario* consumers tend to reduce their service level which is expressed by decreasing mass and engine power of cars sold. This effect together with the diffusion of highly efficient propulsion systems like Plug-In-Hybrids and electric cars causes a reduction of average emissions (see Figure 7-23). The emissions are compared on a well-to-wheel basis (without vehicle production), considering fossil pathways for both internal combustion engine cars (gasoline, diesel & CNG) and electric cars (electricity from natural-gas-fired gas and steam turbines). In the *BAU & Low Price Scenario* the average GHG emissions of cars sold decrease from 180 g km^{-1} to 140 g km^{-1} up to 2030. In the *Policy Scenario & Low Price* a substantial reduction is achieved with average emissions of around 110g km^{-1} in 2030.

At conditions of high fossil fuel prices (High Price Scenarios) mass and power decrease even in the BAU scenario leading to lower average greenhouse gas emissions (120 g $km^{-1)}$. In the *Policy Scenario + High Price* the average emissions decrease to less than 100 g km^{-1}.

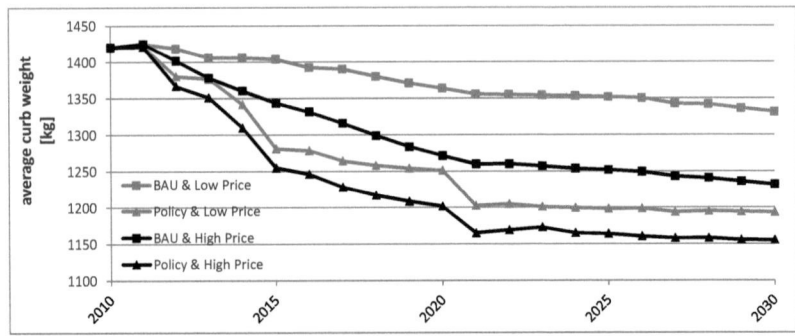

Figure 7-21: Development of average curb weight of new cars in the four main scenarios

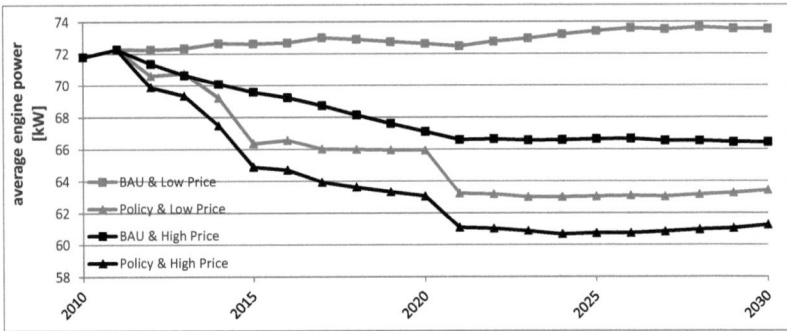

Figure 7-22: Development of average engine power of new cars in the four main scenarios

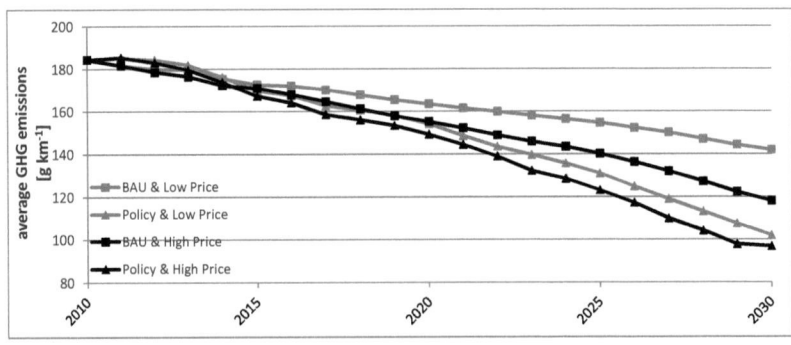

Figure 7-23: Development of average greenhouse gas emissions of new cars in the four main scenarios (WTW without vehicle production)

8 Energy consumption, energy carriers and greenhouse gas emissions of the passenger car fleet in Austria 2010-2050

8.1 Final energy demand in the scenarios (TTW)

In the scenario results final energy demand is broken down in five main energy carriers considered in the model: Gasoline, Diesel, CNG, electricity and hydrogen (the latter is not considered in this demonstrated scenarios). Balances of final energy consumption only show the TTW part of the energy conversion chain and therefore give an incomplete view of the real primary energy consumption in the scenarios. Especially with biofuels and electricity, where major energy conversion losses occur in the fuel production (WTT), the results might be misleading. In the given analysis the final energy consumption served as a basis for a more detailed view on fuel sources and the corresponding energy conversion pathways that will be demonstrated in the following chapters.

The development of final energy demand in the analyzed scenarios indicates the strong effect of political framework conditions and fossil energy prices on the energy consumption and the energy carrier-mix.

In the BAU scenario at low energy prices the diffusion of hybrid cars, slows down the demand increase of final energy carriers (see Figure 8-1) but final energy demand still keeps growing until about 2030 (+14 %). The energy carrier mix will remain dominated by gasoline and diesel fuels. Electricity plays a minor role even in a long run. The dashed lines indicate the sensitivity of the results to changes in the learning parameter used for the battery cost development. It shows that variations in the learning parameter have an impact on the development of the final energy consumption in the *BAU Scenario*. A higher learning rate leads to a stronger diffusion of electric cars and thereby to a reduction of final energy consumption in a long-term.

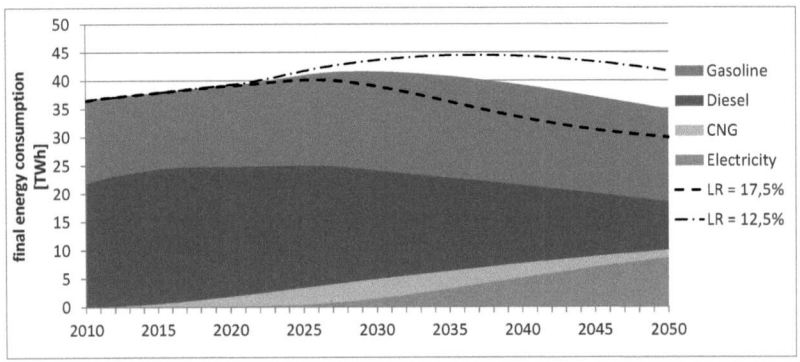

Figure 8-1: Final energy consumption and energy carriers: *BAU & Low Price – Scenario*

The final energy consumption in the *Policy Scenario* is decreasing by about 45% up to 2050 (-32 % up to 2030) (see Figure 8-2). This development is driven by two factors: Firstly, the higher price level leads to a lower yearly kilometrage of the entire fleet and secondly the new vehicles are smaller and use more efficient technologies. These effects even compensate the additional demand caused by the growing car fleet.

In the *Policy-Scenario* the diffusion of electric vehicles leads to a growing importance of electricity as energy carrier. In 2050 the electricity demand of the passenger car fleet reaches 13.6 TWh which is more than 60 % of final energy consumption. In this scenario the demand for conventional fuels like gasoline and diesel decreases by almost 80 %. In this scenario the learning parameter has much less impact on the long term development of final energy consumption than in the BAU scenario.

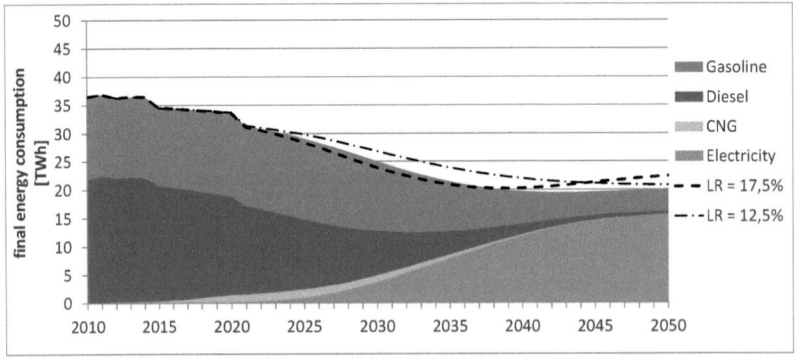

Figure 8-2: Final energy consumption and energy carriers: *Policy & Low Price – Scenario*

The comparison of all analyzed scenarios depicted in Figure 8-3 highlights the effects of policy measures on the energy demand of the fleet which is significantly stronger than the effects of oil price driven fuel price increases.

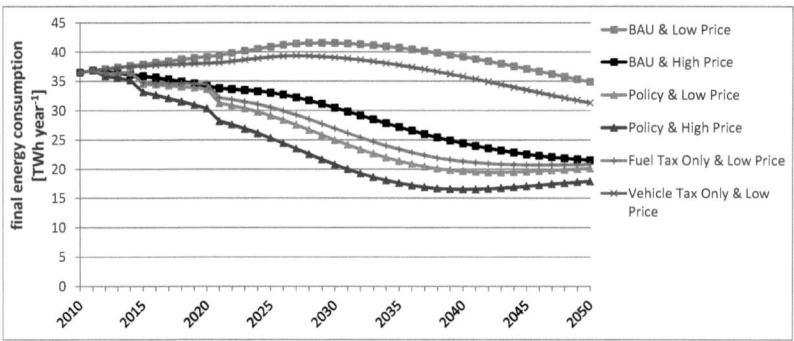

Figure 8-3: Final energy consumption of analysed scenarios

8.2 Fuels and fuel sources 2010-2050

In the model all five fuel types are linked to specific vehicle technologies that can be chosen by the consumers. These fuels are gasoline, diesel, compressed natural gas (CNG), electricity and hydrogen.

These fuels can be produced from conventional fossil sources such as crude oil or natural gas, but they can also contain fractions of alternative fuels based on renewable sources. A detailed view on the considered fuels and their corresponding sources are given in Table 8-1. In Austria there are obligatory rates for biofuel blending of 5.75 % set by the EU biofuel directive. This rate has been met in Austria since 2008 by blending diesel with biodiesel and gasoline with first generation bioethanol (Winter 2008). Contrary to other countries (e.g. Brazil) pure biofuels are hardly available since capacities are all used for large scale blending. For that reason the use of pure fuels are neglected as option for consumers in the model and it is assumed that the demand for biofuels is determined by the overall demand of fuels and by the blending rate set by blending regulation.

The main fuels considered in the analysis can be produced from different fossil or renewable sources (see Table 8-1). Every fuel source with its corresponding energy conversion pathway has a different energy balance and different greenhouse gas emissions. In a well-to-wheel assessment the detailed composition of the fuels and fuel sources has to be considered.

Gasoline: in Austria gasoline is still mainly based on fossil sources. Since 2008 gasoline is blended with 5.75 % of ethanol produced from wheat, corn and sugar beet (Winter 2008). This type of ethanol is also called first generation ethanol *(ethanol 1)*. In the future it is expected that first generation ethanol is going to be complemented with second generation ethanol *(ethanol 2)*. In second generation ethanol the entire plant can be used to produce the fuel instead of only the fruits or seeds. Therefore, they need less land to produce the same amount of energy and have a significantly better greenhouse gas balance (see chapter 8.2.1). Furthermore, they permit a better diversification of fuel sources.

Diesel: today diesel is blended with 5.75 % of biodiesel in Austria, which is mainly produced from rapeseed and sunflowers (Winter 2008). Just like *ethanol 1* it is expected that biodiesel is going to be substituted by more advanced second generation fuels in the next years and decades. One of these second generation fuels is *Fischer-Tropsch diesel* that can be produced from cellulose or lignocelluloses and offers the same advantages as second generation ethanol.

Compressed natural gas (CNG): CNG plays a minor role as transport fuel in Austria. CNG is almost exclusively based on fossil natural gas. Today there are only a few facilities that feed in biogas in the natural gas grid in Austria and the quantities are insignificant (Winter 2008). In the upcoming decades the share of fed-in biogas is expected to increase. The main sources for biogas are manure, maize silage and different types of energy plants. Another potential future blend of natural gas is synthetic natural gas (SNG) produced from lignocelluloses materials like wood chips.

Electricity: because of its characteristics there are fundamental differences between electricity and the other transport fuels in the entire energy conversion chain. The relevant aspects of electricity as transport fuel will be discussed in detail in chapter 8.3.

Hydrogen: Hydrogen is often treated as the ultimate solution for a future transport fuel. With its high specific energy and zero TTW-emissions hydrogen is definitely a promising fuel for passenger cars from a technical perspective. However, hydrogen is still far away from being used as transport fuel in larger scale today. There are still many technological and economic barriers to overcome to make the fuel a feasible option (see also chapter 5.2.8). Today, it is almost impossible to estimate when these problems are going to be solved and when hydrogen-based passenger car transport is going to become a realistic alternative.

In principle hydrogen fuel cell cars are foreseen as a technology option in the model and can also be included in the scenarios. Due to the technology-specific uncertainties and because of the special focus of this thesis on hybrid and electric cars, hydrogen fuel cell cars are not considered in these particular scenarios.

Table 8-1: Fuels and fuel sources

fuel-type	fuel	fossil	renewable	primary sources
gasoline	fossil gasoline	✓		crude oil
	ethanol 1		✓	corn
			✓	wheat
			✓	sugar beet
	ethanol 2		✓	straw
			✓	short rotation coppice
diesel	fossil diesel	✓		crude oil
	biodiesel		✓	rapeseed
			✓	sunflower
			✓	used cooking oil
	Fischer-Tropsch-Diesel		✓	wood chips
compressed natural gas (CNG)	natural gas	✓		natural gas
	biogas		✓	manure
			✓	energy plants
			✓	maize silage
	Synthetic Natural Gas		✓	wood chips
electricity		✓	✓	electricity-mix Austria
		✓		natural gas (gas & steam)
			✓	hydro
			✓	wind
			✓	photovoltaics
			✓	wood chips
			✓	short rotation coppice
hydrogen	compressed hydrogen	✓		natural gas (gas & steam)
			✓	hydro
			✓	wind
			✓	photovoltaics
			✓	wood chips
			✓	short rotation coppice
			✓	energy plant mix

8.2.1 Energy- and greenhouse gas balances of cars and fuels

For a comprehensive analysis of energy consumption and GHG emissions of passenger cars the entire energy conversion chain has to be considered. The corresponding well-to-wheel balances include the production of the fuel (well-to-tank WTT), the production of the car and the energy conversion in the car (tank-to-wheel TTW) (see chapter 5.3). Figure 8-4 and Figure 8-5 illustrate the WTW greenhouse gas emissions of a diesel car and an electric car. The figures point out why it is important to consider the entire energy conversion chain (WTW).

For cars with internal combustion engines that are using fossil fuels, 80 – 90 % of GHG emissions occur in the TTW-part (Kloess et al. 2009). For cars that are using alternative fuels the WTT part can actually be more important. A good example for this is the electric car that has no TTW emissions, since all emissions occur in the WTT part, where the electricity is generated. Moreover, the figures indicate that the production of the car must not be neglected in the WTW energy balance, as they usually account for at least 10% of life-cycle emissions (assumed car life: 225 000 km).

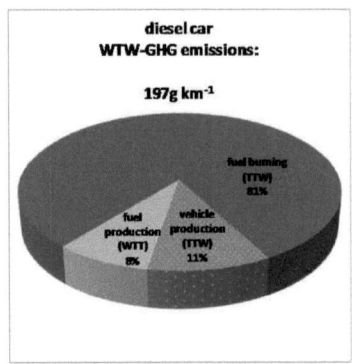
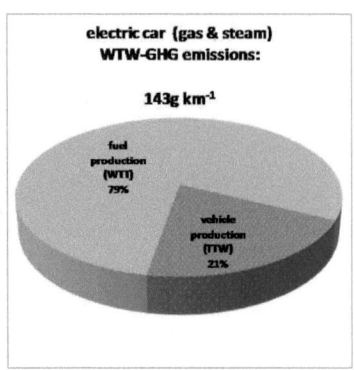

Figure 8-4: WTW greenhouse gas emissions of a diesel car (middle class)

Figure 8-5: WTW greenhouse gas emissions of an electric car (middle class)

The necessary data for all energy conversion pathways was provided by JOANNEUM Research who determined GHG emissions and energy consumption through life cycle analysis LCA. The Life-Cycle-Data includes production and transport of the fuel, the conversion of the fuel in the car and also the embodied energy of the car. The applied methodology of the life cycle analysis is described in (Cherubini et al. 2009) and a more detailed view on the data used in this analysis can be found in (Kloess et al. 2009).

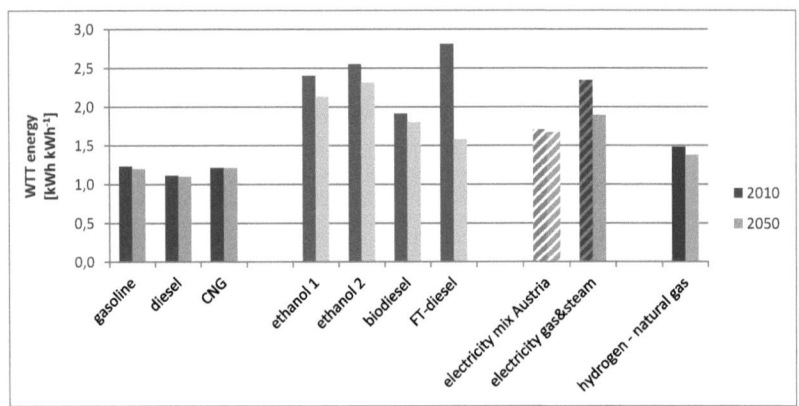

Figure 8-6: WTT energy of selected fossil and renewable fuels (Data Source: Joanneum Research; see (Kloess et al. 2009))

The WTT energy balance of the fuels illustrated in Figure 8-6 gives an overview on the amount of energy that is needed to produce 1 kWh of final energy. It shows that especially biofuels need high amounts of primary energy for their production. However, the major part of this energy comes from renewable sources (organic material). Also electricity production cause considerable losses in the production phase.

Figure 8-7 depicts the well-to-wheel energy balances of different fossil fuels and biofuels and different electricity pathways for the technological status 2010. Thereby, primary energy consumption is split up into fossil and renewable fractions.

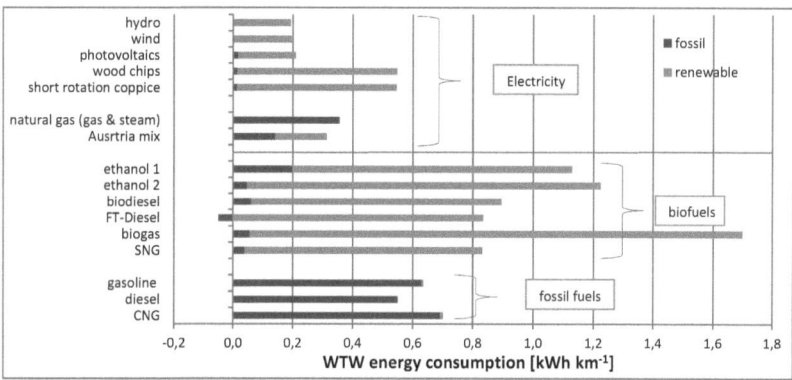

Figure 8-7: WTW energy consumption of selected fossils fuels, biofuels and electricity pathways 2010 (Data Source: Joanneum Research; see (Kloess et al. 2009))

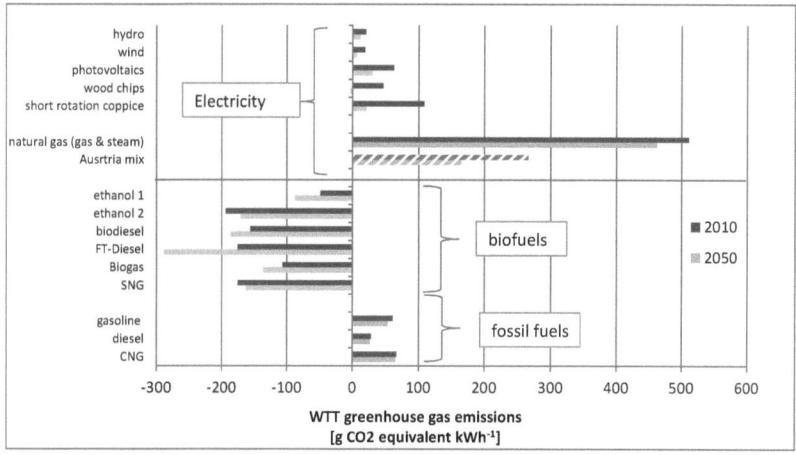

Figure 8-8: WTT greenhouse gas emissions of selected fossils fuels, biofuels and electricity pathways (Data Source: Joanneum Research; see (Kloess et al. 2009))

Figure 8-8 depicts the TTW greenhouse gas emissions of fossil fuels biofuels and different electricity pathways 2010 and 2050. In the TTW phase biofuels have negative emissions since the plants absorb CO_2 during their growth. Fossil fuels like gasoline and diesel also cause emissions in their production and distribution. In the WTT balance the emissions of diesel is considerably lower than the one of gasoline or CNG. The changes in the balances between 2010 and 2050 are caused by the improvement of processes and the changes of the reference system[2].

Figure 8-8 also shows the WTT energy balance of fossil and renewable electricity production pathways and the Austrian supply mix 2010 and 2050. It shows that even pathways with renewable sources of electricity which are often considered CO_2 neutral cause greenhouse emissions in their life-cycle. Due to its high share of hydro-energy the Austrian electricity mix has relatively low emissions today.

The greenhouse gas emissions that are caused during the car production for different propulsion systems in the middle class are depicted in Figure 8-9. It shows that production related emissions increase with increasing complexity of the propulsion system. Especially electrified cars cause considerably higher emissions in their production phase than conventional ones.

The corresponding TTW GHG emissions from the use of the fuel are illustrated in Figure 8-10. It is evident that these emissions are correlated with the fuel consumption and the efficiency of the cars. Electricity and hydrogen based cars have no emission in this balance since no hydrocarbons are burned in the car.

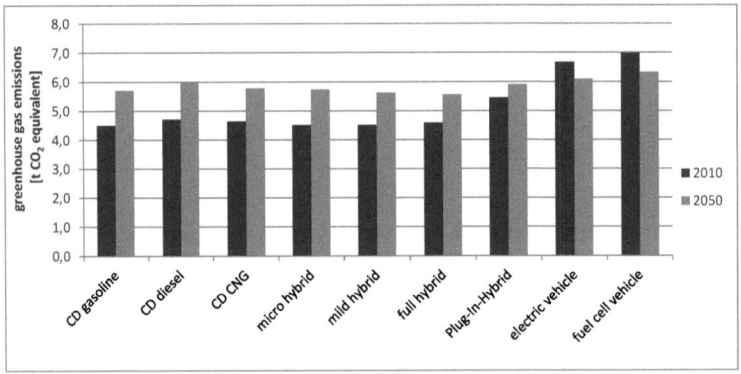

Figure 8-9: greenhouse gas emissions from vehicle production (middle class cars) (Data Source: Joanneum Research; see (Kloess et al. 2009))

---

[2] The reference system is especially relevant in the case of biofuels, where it refers to the emissions that would be caused if the land was not used for biofuel production.

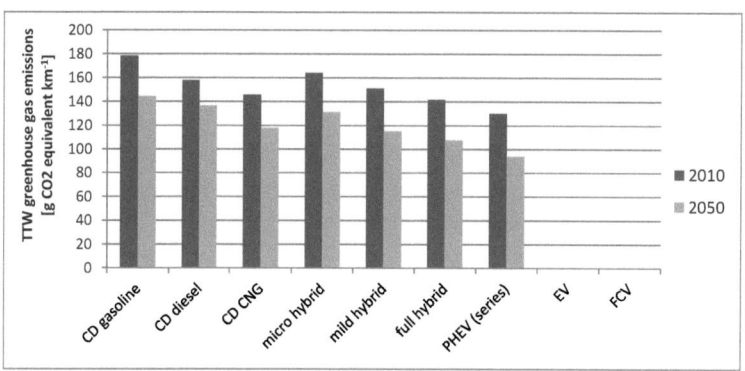

Figure 8-10: TTW greenhouse gas emissions 2010 & 2050 (middle class cars) (Data Source: Joanneum Research; see (Kloess et al. 2009))

Figure 8-11 shows the WTW GHG emissions of selected fuels applied in middle class cars, comparing conventional drive cars with gasoline respectively ethanol 1 or 2 and diesel respectively Biodiesel and FT Diesel, with an electric car and a fuel cell car. It demonstrates that emissions can be reduced by using first generation biofuels like ethanol 1 or biodiesel. However, a considerable reduction requires the use of second generation biofuels like ethanol 2 or FT diesel. Also electricity based pathways show considerably lower emissions even if electricity is produced from fossil sources like natural gas. With its high share of renewable sources the Austrian electricity supply mix would permit WTW emissions of less than 90 g km^{-1} for middle class cars which is less than 50 % of a diesel car's WTW emissions.

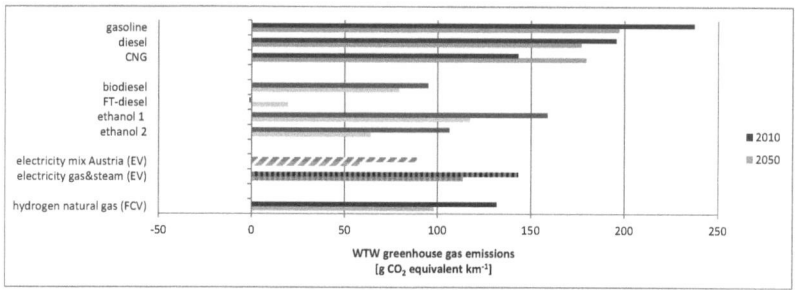

Figure 8-11: WTW greenhouse gas emissions of selected conversion chains (middle class cars 15 000km year^{-1}) (Data Source: Joanneum Research; see (Kloess et al. 2009))

8.2.2 Scenarios for biofuel blending

In long-term scenarios (e.g. 2010-2050) the sources for fuels are likely to change due to changing economic, technological and regulative framework conditions. In the model it is assumed that the sources of the blending fractions continuously shift to more advanced options. For example first generation biofuels (Biodiesel, Ethanol 1) get substituted by more efficient and environmentally compatible second generation biofuels (FT Diesel, Ethanol 2).

The percentage of biofuel blends used in transportation fuels depends on the specific policy in the country or region. Following the EU biofuel directive Austria has a 5.75 % biofuel fraction in both gasoline and diesel fuels (Winter 2008). In the analysis two different scenarios of biofuel blending for the timeframe 2010-2050 are analysed (see Table 8-2).

Business-As-Usual biofuel Policy: in this policy scheme it is assumed that efforts are made to meet the EU biofuel directive for 2020 by raising the biofuel blending rate to 8.5 %.

Active biofuel policy: in this case an ambitious biofuel policy is assumed. After reaching the 2020 biofuel blending goals the blending rate is further increased up to a share of 30 % in 2050.

In the model the policies concerning biofuel blending are considered as independent from the above mentioned policies of car- and fuel taxation. In practice high shares of biofuels would have an effect on the price of transport fuels. In other words the active biofuel blending policy is more likely to be implemented in a high fossil fuel price scenario.

Table 8-2: Scenarios of biofuel blending

	2010	2020	2030	2040	2050
Business as usual biofuel policy	5,75%	8,50%	8,50%	8,50%	8,50%
Active biofuel policy	5,75%	10%	20%	25%	30%

Fort the amounts of biofuels that are required for the different blending rates in the scenarios see Appendix B.

8.3 Electricity as fuel for passenger cars

Electricity as fuel for passenger cars fundamentally changes the way cars are refuelled. Electric cars are not charged on one central fuel station like conventional cars, but distributed at the homes of their owners or at parking lots. Unlike conventional cars that have to be refuelled once a week or even more rarely electric cars have to be recharged every day due to their low driving ranges. When they are plugged in and they are charging they become an additional load in the electricity grid that has to be covered by additional electricity supply.

8.3.1 Effects of Electric Vehicle Charging on the load profile

In this chapter the effects of cumulative EV charging on the Austrian electricity load profile will be estimated. The time the vehicles are charging via the grid can be derived from the typical user profile of passenger cars in Austria (see (Litzlbauer 2009)). Here it has to be differentiated between controlled and uncontrolled charging. In the case of uncontrolled charging electric cars start charging whenever their users plug them in. For a typical user this will happen at the end of a trip or at the end of the day depending on the availability of infrastructure. The most common way consumers will charge their PHEVs or EVs will be at their homes after returning from work. There are fears that the concurrent charging of huge numbers of EVs could cause critical load peaks in the grid. Theses load effects have been analysed for the derived fleet penetration scenarios.

8.3.1.1 Cumulative Load Profile of Electric Cars

Figure 8-12 illustrates the theoretic case that all vehicles in the fleet in 2030 (*Policy + Low Price Scenario*) plug in at the same time on a household plug (230 V; 16 A) and keep charging until their battery is entirely recharged after driving their average daily driving distance (see equation (8-1)) (for detailed data see Appendix B

$$P_{cum\_max\_theor} = CAP_{EV} \cdot P_{plug} \quad [GW] \tag{8-1}$$

$P_{cum\_max\_theor}$ ... *theoretic maximum load of electric vehicles [GW]*
CAP_{EV} ... *electric vehicles in the fleet (BEVs & PHEVs)*
P_{plug} ... *charging power [GW vehicle^{-1}]*

In 2030 this would cause an extra load of more than 6 GW which is about 75 % of the peak load caused by final consumers on a winter day in Austria in 2010 (cf. (e-control 2010a)). This theoretic case of simultaneous charging would never occur in practice. Due to the individual user profiles of cars in the fleet there will be certain distribution of the recharging processes during the day. In the following analysis the cumulative load profile of electric vehicles in 2030 is estimated using a stochastic approach.

In a first case it is assumed that the cumulative load caused by electric vehicles is normally distributed around the expectancy value 18 h, representing the case that vehicles are mainly charged at home. In addition there is the case that assumes that a certain share of cars can be charged at work, causing another load-peak in the morning and flattening the evening peak. Also the morning peak is approximated by a normal distribution with an expectancy value at 8 h (see equations (8-2) and (8-3) and Figure 8-14). Figure 8-15 gives the cumulative load profile of all BEVs and PHEVs in the *Policy & Low Price - Scenario* in 2030 that leads to the daily electricity consumption defined by equation (8-4). In the morning and evening charging-case it is assumed that 80 % of the energy for EVs is charged in the evening and 20 % in the morning. In the other case it is assumed that EVs are only charged in the evening (100%).

$$P_{cum\_EV}(x_r) = z_m \cdot f(x_r, \mu_m, \sigma_m) + z_e \cdot f(x_r, \mu_e, \sigma_e) \quad [GW] \quad (8\text{-}2)$$

$$f_{(x)} = \frac{1}{\sigma\sqrt{2\pi}} \cdot e^{-\frac{1}{2}\left(\frac{x-\mu}{\sigma}\right)^2} \quad (8\text{-}3)$$

P_{cum_EV} ... cumulative load caused by electric vehicles charging [GW]
z_m ... percentage of electricity charged in the morning [%]
z_e ... percentage of electricity charged in the evening [%]
x_r ... discrete random variable (= hour of the day)
µ ... expectancy value of the load peak (m ... morning; e ... evening)
σ ... standard deviation if the load peak (m ... morning; e ... evening)

$$E_{EV} = \sum_{n,i,j,u} d_{EV} \cdot EC_{EV} \cdot \eta_{cg}^{-1} \quad [GWh] \quad (8\text{-}4)$$

$$E_{EV} = E_{EV} \cdot \left(z_m \cdot \int_0^{24} f(x_r, \mu_m, \sigma_m) dx + z_e \cdot \int_0^{24} f(x_r, \mu_e, \sigma_e) dx \right) \quad [GWh] \quad (8\text{-}5)$$

E_{EV} ... cumulative daily electricity consumption [GWh]
d_{EV} ... distance driven in electric mode per day(BEVs & PHEVs) [km]
EC_{EV} ... electricity consumption of EVs and PHEVs in electric driving mode [Wh km^{-1}]
η_{cg} ... charging efficiency of electric vehicles [%]

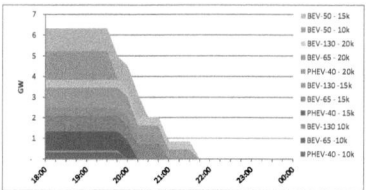

Figure 8-12: Theoretic load caused by simultaneous charging of all BEVs and EVs 2030 (*Policy & Low Price Scenario*)

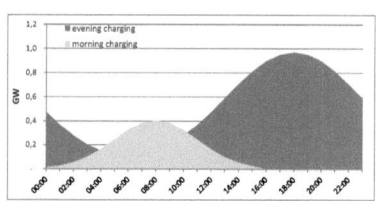

Figure 8-14: Approximation of the daily load profile of all EVs 2030 – 80% evening & 20% morning charging

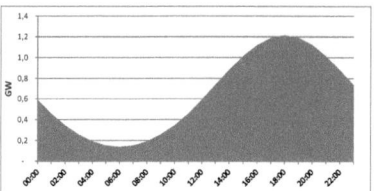

Figure 8-13: Approximation of the daily load profile of all EVs 2030 – evening charging only

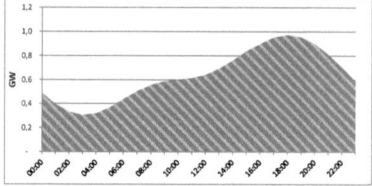

Figure 8-15: Cumulative load profile of EVs 2030 – evening & morning charging

Figure 8-16 to Figure 8-19 illustrate the impact of EVs (BEV & PHEV) on the electricity load profile in Austria in 2030, 2040 and 2050. Thereby, the load caused by the EVs is added to selected load profiles of end user consumption for a winter and a summer day in Austria 2010, taken from (e-control 2010a). All figures show the case of morning and evening charging (red) as well as the case of evening charging only (dashed line). Figure 8-16 depicts the winter load curve resulting from the 2030-EV-diffusion and Figure 8-18 depicts the corresponding curve for a summer day.

With the EV fleet penetration in 2030 (36 % of the fleet) the additional load on the evening peak (18 h) is 0.9 (morning & evening charging) to 1.2 GW (evening charging only). In 2050 with an EV fleet penetration of 96 % the additional load accounts for 2.8 to 3.5 GW.

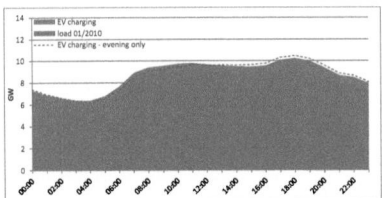

Figure 8-16: Effect of EV charging on the load profile 2030 – winter day

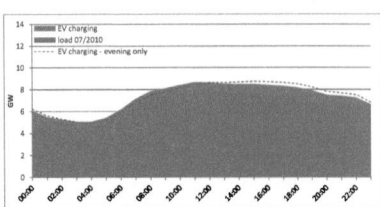

Figure 8-18: Effect of EV charging on the load profile 2030 – summer day

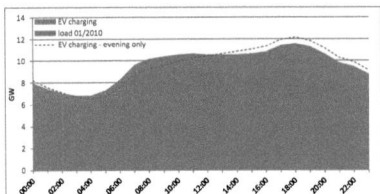

Figure 8-17: Effect of EV charging on the load profile 2040 – winter day

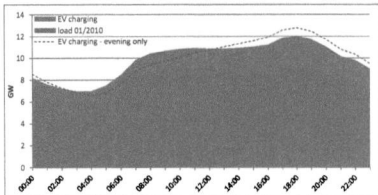

Figure 8-19: Effect of EV charging on the load profile 2050 – winter day

The results indicate that shifting part of the charging in the morning hours could to some extent mitigate the critical load peaks in the evening hours. The question whether there is only "evening charging" or "morning and evening charging" is mainly a question of infrastructure availability. Only if sufficient public charging infrastructure is available (for example at the work place) cars can be recharged during the day.

Even if morning and evening charging is possible the load caused by EVs still adds a critical extra load that will have to be covered with peak load power plants. In order to shift some of the charging to lower load times controlled charging has to be applied. This so called grid-to-vehicle (G2V) concept can help shifting load caused by EVs from peak to off peak times. In a further step there are even concepts that consider electric cars as potential electricity storage capacities that could be used to store electricity in off peak times and to feed-in in peak times (vehicle-to-grid V2G).

8.3.2 Electricity supply scenarios

When analysing electricity based passenger transport it is important to differentiate between generation technologies and their corresponding primary energy sources. Their emissions and primary energy consumption can be very different which significantly affects the overall energy and greenhouse gas balances. For an assessment of these balances the production mix of the electricity has to be known in detail. To exactly determine the production mix of the electricity that the cars receive, a detailed analysis of the load profile as well as the production mix would have to be performed (see Figure 8-20).

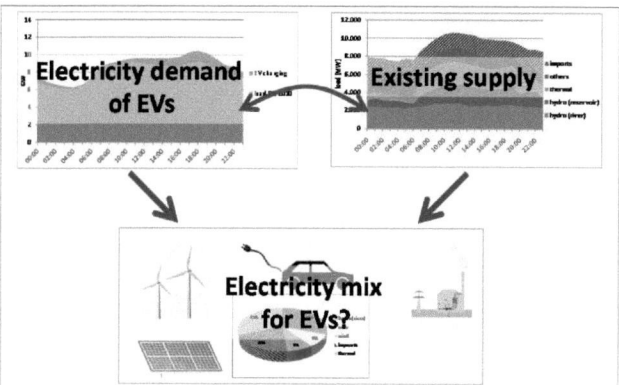

Figure 8-20: electricity production mix for electric cars

To determine yearly greenhouse gas emissions in the scenarios 2010-2050 predetermined yearly electricity supply-mixes are used. In this case there remains the uncertainty which generation capacities are to be considered. In principle there are two interpretations of this issue: There is one that argues that the existing mix has to be considered since there would be no capacities that are exclusively reserved for electric cars. According to this approach average emissions of the Austrian electricity mix would have to be used in this analysis (respectively the projected electricity mix 2010-2050).

The other approach says that the capacity of the existing mix is already used which means that only additional capacities can be considered for EV charging. In this case the emissions of these marginal capacities would have to be considered.

In this specific analysis it is assumed that in a short term, as the required capacity is practically negligible, the supply would be based on the Austrian mix. In a mid to long term an increasing share of the supply for EVs will have to be covered by additional capacities that can either be based on fossil or renewable sources. Consequently two electricity supply scenarios were determined (see table Table 8-3):

Electricity "Fossil": In this scenario it is assumed that the electricity supply will be based on the Austrian mix first and as demand increases it will be complemented with fossil electricity from natural gas fired gas and steam plants.

Electricity "Renewable": In this scenario the supply for EVs will be shifted to pure renewable sources. The mix implies a high share of de-central supply with shares of photovoltaic, small hydro, wind and biomass (see Table 8-3).

Table 8-3: electricity supply scenarios

		2010	2020	2030	2040	2050
"Fossil" supply scenario						
electricity-mix Austria		100%	75%	50%	25%	0%
fossil sources	natural gas (gas & steam)		25%	50%	75%	100%
renewable sources	hydro					
	wind					
	photovoltaics					
	biomass					
"Renewable" supply scenario						
electricity-mix Austria		100%	75%	50%	25%	0%
fossil source	natural gas (gas & steam)					
renewable sources	hydro		10%	25%	25%	25%
	wind		7,5%	10%	20%	25%
	photovoltaics		2,5%	10%	20%	25%
	biomass		5%	5%	10%	25%

8.3.3 Required electricity in the scenarios

This chapter analyses the electricity demand caused by electric cars in the analysed scenarios. Figure 8-21 depicts the electricity demand of electric cars in the four main scenarios. The results indicate that between 2010 and 2020 the electricity demand caused by electric cars will be negligible, especially when having in mind that the overall electricity consumption in Austria was 65.6 TWh in 2009 (entsoe 2010). Starting in 2020 electricity demand is growing in all scenarios. In 2030 yearly electricity consumption ranges from 1.5 TWh *(BAU & Low Price - Scenario)* to 4 TWh *(Policy & High Price - Scenario)*. Up to 2050 yearly electricity consumption through EV-charging grows to 8.6 TWh *(BAU & Low Price - Scenario)* respectively 15.6 TWh *(Policy Scenario & High Price - Scenario)*

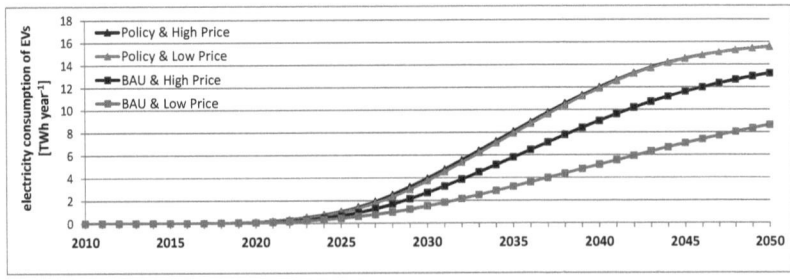

Figure 8-21: electricity demand in the four main scenarios

The derived electricity consumption and the required capacities to cover the demand of EVs in the *Policy & Low Price Scenario* are illustrated in Figure 8-22 and Figure 8-23. The necessary production capacities of gas and steam plants would be 2.2 GW up to 2050 (assuming 6000 full load hours/year)

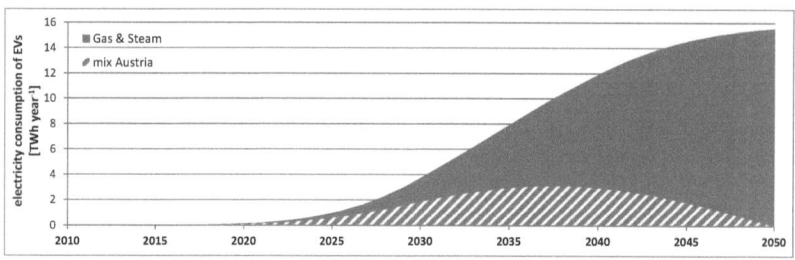

Figure 8-22: electricity demand in the *Policy Scenario* with *fossil electricity* supply

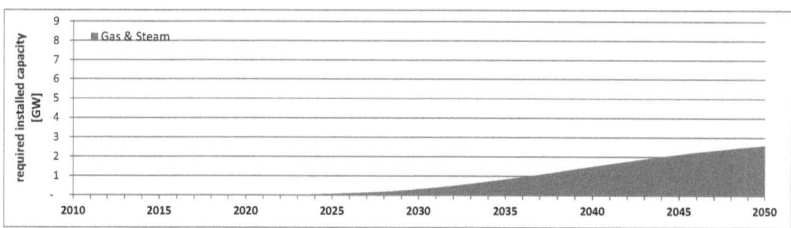

Figure 8-23: required generation capacity in the *Policy Scenario* with *fossil* electricity supply

In order to minimize primary energy demand and greenhouse gas emissions the electricity for EVs has to be generated from renewable sources. This is assumed in the "renewable" electricity supply scenario, where the demand is covered with a mix of renewable sources. The resulting electricity demand split into the four renewable sources is depicted in Figure 8-24 for the *Policy + Low Price Scenario*. Figure 8-25 depicts the generation capacity necessary to provide the electricity required in this scenario. To calculate the capacity typical full load hours for these generation technologies in Austria are used (wind: 2000 h; hydro (small): 4900 h; PV: 900 h; biomass: 6000 h) (Pöppl et al. 2009).

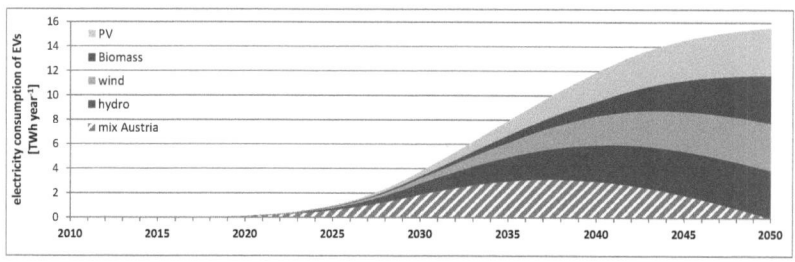

Figure 8-24: electricity demand in the *Policy Scenario* with *renewable electricity* supply

Covering the electricity demand of electric cars in the Policy scenario with the defined renewable electricity mix would require additional renewable capacities of almost 7.6 GW up to 2050 (PV: 4.3 GW; wind: 1.9 GW; hydro: 0.8 GW; biomass: 0.6 GW). Projections of techno-economic feasible renewable electricity potentials show that these capacities can be installed by that time (Ragwitz et al. 2009).

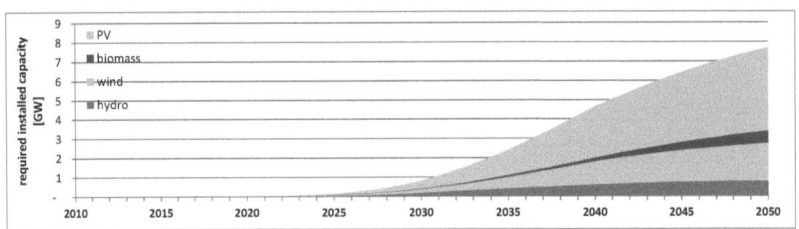

Figure 8-25: required generation capacity in the *Policy Scenario* with *renewable electricity* supply

8.4 Well-to-Wheel (WTW) energy demand and greenhouse gas emissions

To get a comprehensive view on energy consumption and GHG emissions of the passenger car fleet in the scenarios WTW balances are used. The WTW energy balances include the entire energy conversion chain and are broken down into renewable and fossil shares. The WTW greenhouse gas emissions are broken down into emission from fuel production, fuel use and car production (see chapter 8.2.1).

8.4.1 BAU scenario

The results of the BAU scenario show that adoption of hybrid technology alone cannot compensate the increase in energy demand caused by the growing fleet. Figure 8-26 depicts the WTW energy demand in the *BAU* scenario at a low fossil fuel price increase, split up in fossil and renewable fractions. There is an increase of energy demand between 2010 and 2030, and a saturation starting around 2030. The saturation in the long-term is mainly caused by the beginning spread of electric propulsion technologies. With their significantly higher efficiency, they can compensate the effects of increasing demand on energy consumption. The renewable share in the energy balance mainly comes from the blending of biofuels with diesel and gasoline.

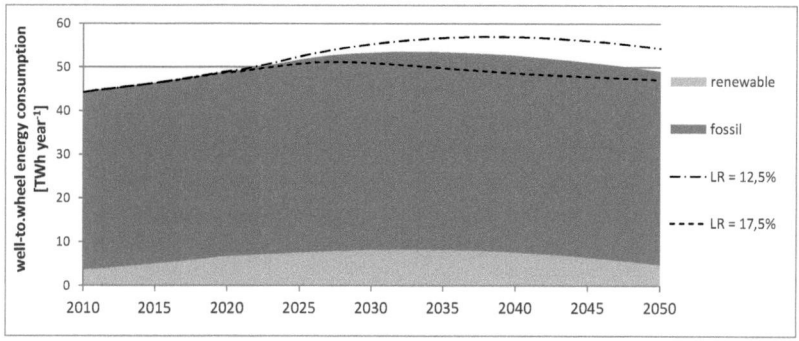

Figure 8-26: **WTW energy demand per year: *BAU & Low Price – Scenario***

Figure 8-27 depicts the well-to-wheel (WTW) greenhouse gas emissions of the entire car fleet caused by burning of the fuel, fuel production and vehicle production for the *BAU scenario* with low fossil energy prices. Similar to the energy consumption emissions keep increasing steadily in the first decades and start to saturate around 2030 (+14 % up to 2030; +15 % up to 2050).

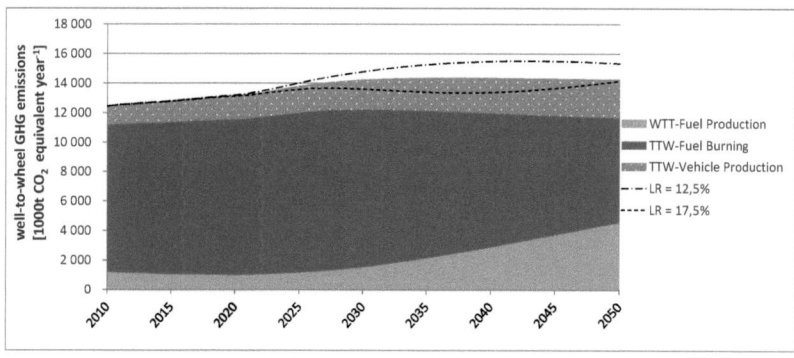

Figure 8-27: **WTW greenhouse gas emissions per year: *BAU & Low Price – Scenario***

8.4.2 Policy Scenario

The results of the *Policy Scenario* give insight on how policy can influence the development of the passenger car fleet in order to reduce greenhouse gas emissions. It demonstrates that a significant reduction of both fossil energy demand and greenhouse gas emissions is achievable through ambitious policy measures in the field.

Figure 8-28 shows the WTW energy balance in the *Policy* scenario at low fossil fuel prices, with *fossil* and *renewable* electricity supply mix. The result clearly demonstrate why it is important to consider WTW data when there are high shares of electricity (or other fuels with high WTT emissions) in the supply mix. When electricity production is considered too energy savings are lower than indicated by the final energy consumption (cf. chapter 8.1). In the *Policy & Low Price scenario* with *fossil* electricity supply total energy consumption is reduced by 21 % up to 2050, which is a respectable reduction bearing in mind that even in this scenario the car fleet keeps growing. In the case of a *renewable* electricity supply the demand for fossil energy can be reduced by more than 86 %.

The results also illustrate the short term effects of higher fuel taxes. The resulting price increase affects the user intensity of all cars, which is expressed by the drops in overall energy consumption in the years where taxes are raised.

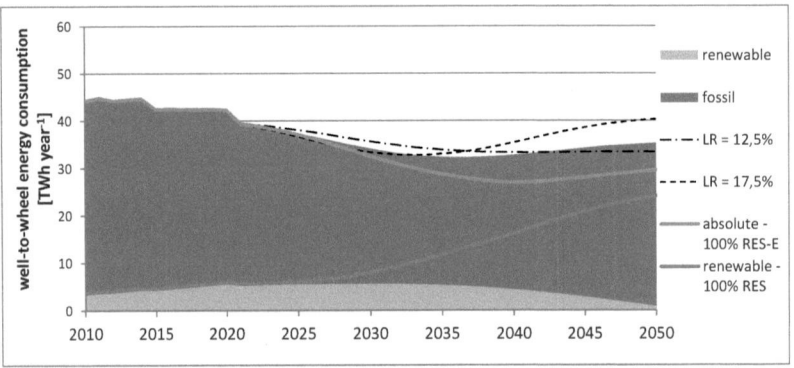

Figure 8-28: WTW energy demand per year: *Policy & Low Price – Scenario*

Slower fleet growth, better efficiency of cars and less carbon intense fuels lead to a considerable reduction of GHG emissions in the *Policy-Scenario*. Figure 8-29 depicts the corresponding WTW greenhouse gas balance for the fossil supply scenario and the renewable supply scenario (100% RES-E electricity). Driven by the growing demand for electricity, emissions from fuel production increase. Also, emissions from vehicle production increase because of the higher shares of electrified cars that cause higher emissions in their production.

With *fossil* electricity GHG emission decline by 26 % up to 2030 and start to increase after 2035. That points out the necessity of higher taxes on electric cars after 2030 to counteract the rebound effect in this scenario. When the electricity supply is shifted toward renewable sources the WTW-emissions can be reduced by 33 % up to 2030 and by 68 % up to 2050. This points out that the full GHG reduction potential of electrified cars can only be tapped if the electricity mix is shifted to low carbon sources.

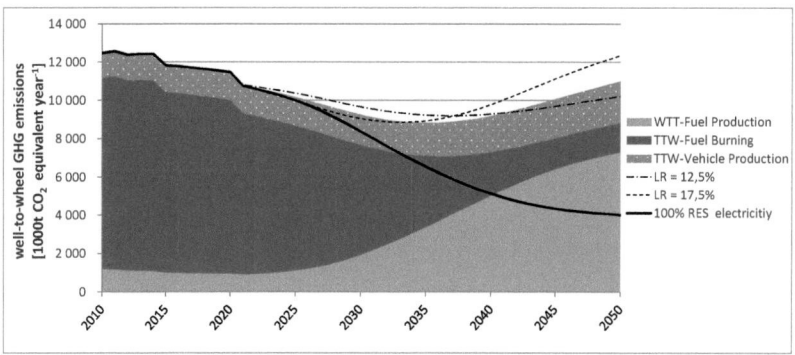

Figure 8-29: **WTW greenhouse gas emissions per year: *Policy & Low Price – Scenario***

8.4.3 Scenario comparison

In Figure 8-30 and Figure 8-31 the WTW energy and the fossil WTW energy consumption of all analysed scenarios are compared. They show that considerable reductions of energy consumption and greenhouse gas emissions are mainly achieved in the *Policy scenarios* (triangles). Even though higher fossil fuel prices also lead to a reduction in the *BAU-Scenario*, the effect in the *Policy scenarios* is significantly stronger.

Enforced use of biofuels can help to reduce fossil energy consumption but they increase the aggregate energy consumption driven by the higher input of energy intense biomass. The reductions in fossil energy consumption that are achieved in the Policy scenarios range from -16 % *(Policy & Low Price + fossil electricity)* to -92 % *(Policy & High Price + renewable electricity)* up to 2050. This shows the key role of the electricity supply mix for fossil energy consumption. In the *Policy Scenarios* with *fossil* electricity supply fossil energy demand re-increases after 2035. This is mainly because of the increased transport demand (fleet growth + higher car use + higher service level) which indicates that electric cars can also implicate rebound effects on energy demand.

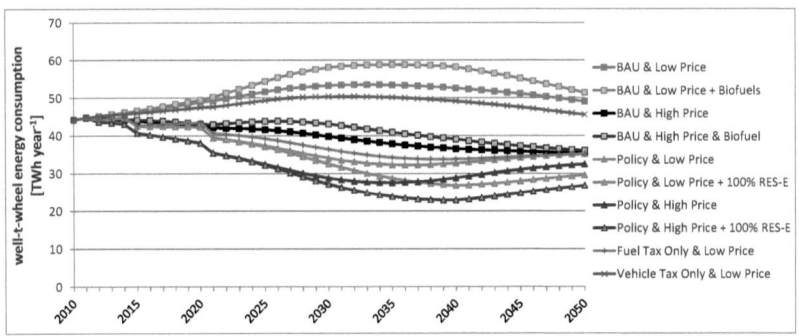

Figure 8-30: WTW energy consumption (embodied energy) of the passenger car fleet in all scenarios

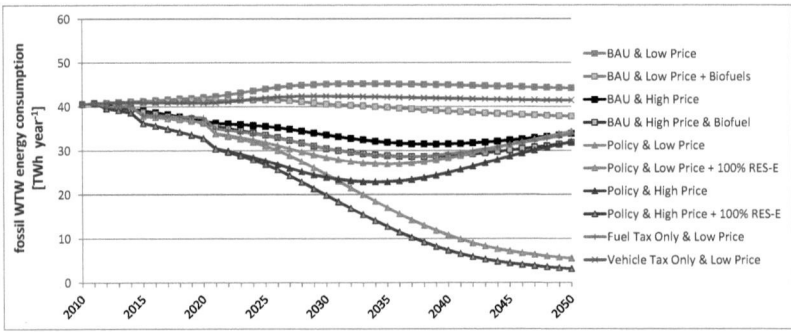

Figure 8-31: Fossil WTW energy demand of the passenger car fleet in all scenarios

The WTW GHG balances of all scenarios depicted in Figure 8-32 also illustrate the impact of policy measures. In the BAU scenario GHG emissions keep increasing (*BAU & Low Price*) when no additional measures are taken. Higher biofuel fractions lead to a stabilisation of emission at the 2010 level (*BAU & High Price + biofuel*). Higher fossil fuel prices lead to a slight reduction of GHG emissions even in the BAU case (*BAU & high price*). However, considerable reductions are only achieved in the *Policy scenarios* (triangles). The reductions of greenhouse gas emissions compared to 2010 achieved in the Policy scenarios range from -14 % (*Policy & Low Price – fossil electricity*) to -75 % (*Policy & High Price – 100% RES-E*). The broad range of reductions in the policy scenarios is mainly caused by the differences in the assumed electricity supply mix. In order to maximize the GHG benefit from electric cars they have to be supplied with low carbon electricity.

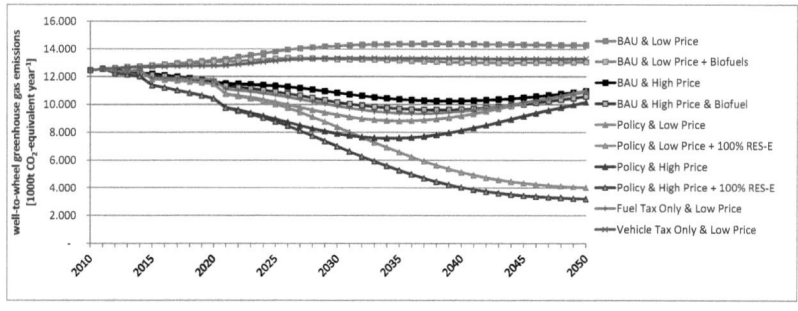

Figure 8-32: **WTW greenhouse gas emissions in all scenarios**

9 Conclusions

The final chapter will draw conclusions on the key findings of this thesis focussing on the two main parts: techno-economic assessment and model-based analysis.

9.1 Techno-Economic Assessment

From a pure technical point of view electrification/hybridisation of the powertrain is an effective measure to cut energy consumption and greenhouse gas emissions of passenger cars. However, electrification also means higher investment costs of cars. The economic optimal degree of electrification depends on the costs of key components and the specific economic and political framework conditions.

- With today's (2010) costs of electric components, gasoline prices have to be higher than 1.5 € liter^{-1} for hybrid systems to become cost effective for a broad range of consumers. Below this price level only micro hybrid systems can compete with conventional technology at average annual driving distances.

Fully electrified propulsion technologies like plug-in hybrid electric vehicles (PHEV) need gasoline prices higher than 2 € liter^{-1} to be cost effective. This points out that the costs of pure electric propulsion systems are still too high with batteries as the main cost drivers. To become economically competitive with conventional cars they will rely on a reduction of battery costs and increasing gasoline and diesel prices.

- Batteries are the key components for vehicle powertrain electrification. They are the main cost driver of electric cars and the success of these cars will strongly depend on the specific cost and the technical reliability and durability of battery systems. In the last two decades batteries have improved considerably and today they meet the technical and economic criteria for hybrid cars. However, fully electric propulsion systems need further improvement in technical performance and considerable reduction in specific costs of batteries. For PHEVs and battery electric vehicles (BEV) to become cost effective specific battery cost (2010: ≈700 € kWh^{-1}) have to be reduced by more than 50 %. The estimation of future battery costs shows that learning effects could lead to the required reduction between 2015 and 2020, if the trend toward hybrid and electric cars continues on a global level.

- The limited driving range and the long recharging time are competitive disadvantages for battery electric cars that are not captured in the cost assessment. Even with today's most advanced battery technology these problems remain unsolved. However, performance of battery technologies is expected to keep improving bringing electric cars closer to consumer requirements in terms of driving range.

One approach to achieve comparable driving ranges with electric cars today is by using range extenders that could be based on internal combustion engines or even on hydrogen fuel cells. The question whether pure electric cars or range extender cars and plug-in hybrids will make it to the mass market will be rather a question of consumer acceptance than cost.

- The assessment of hydrogen fuel cell propulsion systems have shown that fuel cell cost have to come down to less than 200 € kW^{-1} in order to compete with other zero emission technologies (BEVs & PHEVs).

It will be difficult for fuel cell systems to achieve the necessary cost reduction in a short- to mid-term, especially when considering that there is no bridging technology that could act as technology driver. Unlike battery electric systems which can rely on hybrid technology to help reduce cost of batteries by driving global cumulative production and generate technology spill-overs, there is no such technology link for mobile fuel cell systems. On the other hand fuel cell systems could solve two major problems of electric propulsion systems: storing enough energy on board for long diving ranges and permitting fast refuelling. As long as these problems are not solved with battery systems, hydrogen fuel cells will remain in the play as a long-term option.

- The cost estimation for the time frame 2010-2050 indicates that hybrid systems will be the least cost option in a short term (up to 2020-2025). With a reduction of battery costs and increasing fuel prices PHEVs become the best mid- to long term option for middle class cars. At this condition BEVs will become the first choice for compact class cars whose typical field of application requires lower driving ranges (e.g. urban areas). For both PHEVs and BEVs the economically optimal electric driving range will depend on the specific framework conditions (fuel price + yearly driving range) and the cost of the batteries.

9.2 Model based analysis

The model is developed with the special focus on the analysis of effects of new technologies, fossil fuel prices and policy measures on energy consumption and greenhouse gas emissions in the Austrian passenger car fleet. The derived scenarios give an impression of the dynamics of technological change in the passenger car fleet and the effects of different policy measures on the diffusion of new technologies and fuels as well as the fleet development as a whole. Thereby, the model captures the major factors that affect energy demand of passenger car transport, like fleet growth, characteristics of new cars (mass, engine power, fuel consumption) and use of cars. The time frame 2010-2050 permits to analyse long-term effects of changes in economic and political framework conditions on energy demand and greenhouse gas emissions in the

fleet. This allows policy makers to see the effects of policy options in a wider time horizon, which is especially relevant when long term carbon mitigation goals have to be met.
In the thesis the model is applied to develop four scenarios. The key findings of the scenario results are:

- All scenarios share one major trend: a shift in the passenger car fleet towards hybrid cars. Even though hybridisation will greatly improve the efficiency of the fleet, the results of the **BAU scenario** point out that energy demand and greenhouse gas emissions cannot be reduced by simply switching to hybrid technology.

- In the **BAU scenario** *(& Low Price)*, where no major policy measures are taken WTW energy consumption and greenhouse gas emissions of the car fleet keep growing up to 2030 (WTW-energy consumption: +11 %; WTW-GHG emissions: +14 %). This development is mainly driven by the growth of the car fleet (+27 % up to 2030), high yearly kilometrage and a high service level of cars. The diffusion of more efficient hybrid cars cannot offset the effects of theses drivers in the BAU scenario. Highly efficient fully electric cars (PHEVs & BEVs) only slowly diffuse into the fleet (16 % in 2030) and therefore show little effect on energy consumption and greenhouse gas emissions.

- In the **Policy scenario** higher fuel taxes and higher taxes on inefficient cars lead to a significant reduction of both greenhouse gases and energy demand in the fleet (WTW-energy consumption: -21 % & WTW-GHG emissions: -23 % up to 2030). The higher taxes on fuels and cars leads to a real increase in average transport service cost that leads to a reduction of transport demand and service level. This is reflected in a lower fleet growth (+9 % up to 2030), lower yearly odometer readings and lower average curb weight and engine power of cars registered. Furthermore, the higher fuel prices are a strong driver for the diffusion of PHEVs and BEVs (36 % of the fleet in 2030). By this time the market shares of BEVs will already be 68 %. These high shares can be achieved in this scenario because of an improvement in battery technology (lower cost and better technical performance), but above all because of changing behaviour of car users. There will be more small cars that are mainly used for short distances. For these cars limited driving range won't be such a severe barrier in consumer perception as it is for an average car today.

- The comparison of the **Low Price** and the **High Price** scenarios indicate that higher fossil fuel prices also reduce energy consumption and greenhouse gas emissions by slowing the fleet growth and fostering the spread of efficient propulsion technologies. However, the scenario comparison shows that the policy framework has stronger effects.

- The comparison of the electricity supply scenarios shows that the full potential of greenhouse gas reduction of electric cars can only be exploited with a low carbon electricity supply. The **100 % RES-E** supply scenario shows that a completely decarbonised electricity mix reduces the annual fossil fuel energy demand of the passenger car fleet by 86 %, and greenhouse gas emissions by 68 % up to 2050 *(Policy & Low Price Scenario)*.
- The yearly electricity demand caused by EVs in 2030 accounts for only 6 % (respectively 24 % in 2050) of the 2010 final electricity consumption in Austria. However, cumulative charging of electric cars could cause critical increases of load peaks (+13 % in 2030; +39 % in 2050). To mitigate these peaks controlled charging (grid-to-vehicle) of electric cars has to be applied.

In general the results of the model based scenarios indicate that considerable reduction of GHG emissions and fossil fuel dependence of the passenger car fleet can be achieved by a combination of increased efficiency of cars, lower growth in demand for passenger car transport and a lower average service level of cars. The development of all these parameters can be traced back to the service cost of passenger car transport. Higher transport service costs can slow down growth or even reduce the demand for transport and lower the average service level of cars. Transport service costs can be increased by higher fuel prices (either driven by higher taxes or higher fossil fuel prices) or by higher taxes on cars. An environment of higher fuel prices also fosters the diffusion of electric cars (BEVs and PHEVs). These cars could provide the necessary leap in efficiency to significantly reduce energy consumption and greenhouse gas emissions. Furthermore, they can help to escape the lock-in to fossil fuels and facilitate decarbonisation of the energy supply. It will be one of the future challenges to cover the additional electricity demand of electric cars with domestic renewable electricity sources.

9.3 Outlook

The thesis gives some clear answers concerning the status of electrified cars today and its future potential, but also identifies questions that could be subject to further research.

As the results indicate batteries are the key components for the future success of electrified cars. Estimation of the improvement of their technical performance and the reduction of their cost will therefore remain central questions for analyses related to the future potential of electric mobility.

Another interesting issue is the question of infrastructure requirements and costs. Large scale introduction of electric cars will require the build-up of public charging infrastructure. A detailed economic assessment could help to evaluate to which extent infrastructure could become a barrier to the introduction of electric cars.

The question of infrastructure is also important in the context of long term technology options. Investment in infrastructure can easily turn into sunk costs when new, superior technologies emerge and establish themselves. Or it could lead to a new technology lock-In situation making it difficult for new technologies to diffuse into the fleet. This could be the case with hydrogen fuel cell cars. Even though fuel cell cars today seem farther away from market introduction than battery electric cars they still offer some key advantages over them (e.g. longer driving range, fast refuelling). The fact that both technologies require a new infrastructure could lead to a dilemma which pathway should be developed. In this context the potential role of hydrogen and fuel cells in passenger car electrification deserve closer attention, especially with a time frame up to 2050.

Another interesting issue would be to take a closer look into the energy supply for electrified cars. The results show that the use of electric cars facilitates decarbonisation of passenger car transport. However, this can only be achieved with an electricity supply that is strongly based on renewable sources. In a further step it would be interesting to analyse how additional demand caused by EV charging matches with additional supply from renewable generation on a daily basis. Thereby, it would be possible to determine the electricity mix of electric cars and its corresponding emissions correctly and to analyse the role of controlled charging in this context.

Also the question of fiscal effects on future price of electricity for electric cars deserved a closer examination. Today, taxes on electricity are low compared to taxes on gasoline and diesel, since no fuel tax is imposed. However, fuel tax revenues are an important part of national budget in Austria. Diffusion of electric cars could cause a considerable reduction of tax revenues since less gasoline and diesel would be sold. In order to maintain tax revenues fuel tax will have to be imposed on electricity as well. The consequential increase in electricity price could reduce the economic benefit of electric cars and slow down their diffusion.

10 References

Aberle, G., 2003. *Transportwirtschaft* 4. ed., Oldenbourg Wissenschaftsverlag GmbH.

ACEA, 2010. European Automobile Manufacturers Association - ACEA - Statisitics. Available at: http://www.acea.be/index.php/collection/statistics.

ACEA, 2009. *Tax Guide 2009*, European Automobile Manufacturer's Association. Available at: www.acea.be.

Ajanovic, A., 2008. On the economics of hydrogen from renewable energy sources as an alternative fuel in transport sector in Austria. *International Journal of Hydrogen Energy*, 33(16), pp.4223-4234.

Ajanovic, A. et al., 2009. *ALTERMOTOVE - Country Review Report*, Intelligent Energy – Europe (IEE), STEER Contract no. IEE/07/807/SI2.499569.

Altmann, M. et al., 2004. *Potential for Hydrogen as a Fuel for Transport in the Long Term (2020 - 2030)*, Institute of Prospective Technology Studies ipts.

Andersson, B.A. & Rade, I., 2001. Metal resource constraints for electric-vehicle batteries. *Transportation Research Part D: Transport and Environment*, 6(5), pp.297-324.

Angerer, G., Marscheider-Weidemann, F., Lüllmann, A. et al., 2009. *Rohstoffe für Zukunftstechnologien - Einfluss des branchenspezifischen Rohstoffbedarfs in rohstoffintensiven Zukunftstechnologien auf die zukünftige Rohstoffnachfrage*, Fraunhofer Institut für System- und Innovationsforschung ISI & Institut für Zukunftsstudien und Technologiebewertung IZT gGmbH.

Angerer, G., Marscheider-Weidemann, F., Wendl, M. et al., 2009. *Lithium für Zukunftstechnologien - Nachfrage und Angebot unter besonderer Berücksichtigung der Elektromobilität*, Fraunhofer Institut für System- und Innovationsforschung ISI.

Angerer, G. et al., 2010. *Kupfer für Zukunftstechnologien Nachfrage und Angebot unter besonderer Berücksichtigung der Elektromobilität*, Fraunhofer Institut für System- und Innovationsforschung ISI.

Aral, 2009. *Aral Studie - Trends beim Autokauf 2009*, Aral Aktiengesellschaft.

Autorevue, 2010. *Autokatalog 2010*, Auto Revue Magazin. Available at: http://www.autorevue.at/autokatalog-2010.

Axsen, J., Mountain, D.C. & Jaccard, M., 2009. Combining stated and revealed choice research to simulate the neighbor effect: The case of hybrid-electric vehicles. *Resource and Energy Economics*, 31(3), pp.221-238.

Begluk, S., 2009. *Zukunftsperspektiven der Brennstoffzellenfahrzeuge*, Vienna University of Technology.

Berger, L., Lesemann, M. & Sahr, C., 2009. SuperLight-Car - Innovative Multi Material-Bauweise. *Lightweightdesign*, 9/09, pp.43-49.

Biermann, J., 2008. Elektro Hybrid – Eine Übersicht zu einem Erfolgversprechenden alternativen Fahrzugantrieb. In *Hybrid-, Batterie- und Brennstoffzellenfahrzeuge*. Kontakt und Studium. Expert Verlag.

Braess, H., 1992. Fahrzeugtechnik der Elektro-PKW - Problembereiche, Lösungsmöglichkeiten. In VDI Fachtagung "Elektro Straßenfahrzeuge". VDI Verlag.

Brunner, S. et al., 2010. Der Markt für Elektrofahrzeuge: Politisch-wirtschafliche Rahmenbedingungen der Angebots- und Nachfrageseite. *Energiewirtschaftliche Tagesfragen*, (11/2010), pp.48-53.

Burke, A., Miller, M. & Zhao, H., 2010. Lithium Ion Batteries and Ultra Capacitors alone and in combination in hybrid vehicles: Fuel economy and battery stress reduction advantages. In 25th. World Battery and Electric Vehicle Symposium & Exhibition. Shenzhen.

Button, K., 2010. *Transport Economics* 3. ed., Edward Elgar.

Campanari, S., Manzolini, G. & Garcia de la Iglesia, F., 2009. Energy analysis of electric vehicles using batteries or fuel cells through well-to-wheel driving cycle simulations. *Journal of Power Sources*, 186(2), pp.464-477.

Carlson, E.J. et al., 2005. *Cost Analysis of PEM Fuel Cell NREL/SR-560-39104 Systems for Transportation*, Cambridge: National Renewable Energy Laboratory-NREL.

Ceuster, G.D. et al., 2007. TREMOVE Service contract for the further development and application of the transport and environmental TREMOVE model Lot 1, Brussels: EU Comission DG Environment/Transport & Mobility Leuven.

Chalk, S.G. & Miller, J.F., 2006. Key challenges and recent progress in batteries, fuel cells, and hydrogen storage for clean energy systems. *Journal of Power Sources*, 159(1), pp.73-80.

Chalk, S.G., Miller, J.F. & Wagner, F.W., 2000. Challenges for fuel cells in transport applications. *Journal of Power Sources*, 86(1-2), pp.40-51.

Cherubini, F. et al., 2009. Energy- and greenhouse gas-based LCA of biofuel and bioenergy systems: Key issues, ranges and recommendations. *Resources, Conservation and Recycling*, 53(8), pp.434-447.

Christidis, P., 2003. Trends in vehicle and fuel technologies – review of past trends, EU-JRC, IPTS.

Christidis, P. et al., 2005. Hybrids for road transport - Status and prospects of hybrid technology and the regeneration of energy in road vehicles, EU JRC, IPTS.

Christidis, P., Hidalgo, I. & Soria, A., 2003. *Dynamics of the introduction of new passenger car technologies*, EU JRC, IPTS.

Christl, W., Frickenstein, E. & Scheuerer, K., 1992. Internationale Förderprogramme und Markeinführungshilfen für Elektrostraßenfahrzeuge. In *Elektro-Straßenfahrzeuge*. VDI Berichte. Düsseldorf: VDI Verlag.

Conte, M., Prosini, P.P. & Passerini, S., 2004. Overview of energy/hydrogen storage: state-of-the-art of the technologies and prospects for nanomaterials. *Materials Science and Engineering B*, 108(1-2), pp.2-8.

Cowan, R. & Hultén, S., 1996. Escaping lock-in: The case of the electric vehicle. *Technological Forecasting and Social Change*, 53(1), pp.61-79.

Cuenot, F., 2009. CO2 emissions from new cars and vehicle weight in Europe; How the EU regulation could have been avoided and how to reach it? *Energy Policy*, 37(10), pp.3832-3842.

Dargay, J., Gateley, D. & Sommer, M., 2007. Vehicle Ownership and Income growth, worldwide: 1960-2030. *The Energy Journal*, Volume 28(Nr 4).

Dargay, J. & Gately, D., 1999. Income's effect on car and vehicle ownership, worldwide: 1960-2015. *Transportation Research Part A: Policy and Practice*, 33(2), pp.101-138.

Dreher, M. et al., 1999. Energy price elasticities of energy-service demand for passenger traffic in the Federal Republic of Germany. *Energy*, 24(2), pp.133-140.

e-control, 2010a. Austrian Electricity Operation Statistics. Available at: http://www.e-control.at/de/statistik/strom/betriebsstatistik/betriebsstatistik2010.

e-control, 2010b. Composition of Consumer Electricity Prices in Austria. Available at: http://www.e-control.at/en/consumers/electricity/electricy-prices/price-composition.

entsoe, 2010. electricity consumption data of european countries. Available at: https://www.entsoe.eu/index.php?id=28.

Erdmann, G. & Zweifel, P., 2008. *Energieökonomik*, Springer.

Essen, H.V. et al., 2008. Internalisation measures and policy for the external cost of transport, Delft: CE Delft.

EUCAR, CONCAWE & JRC, E., 2006. Well-to-Wheel Analysis of Future Automotive Fuels and Powertrains in the European Context,

European Commission, 2010. EU Energy and Transport in Figures - Statistical Pocketbook 2010,

European Parliament & European Council, 2003. DIRECTIVE 2003/30/EC OF THE EUROPEAN PARLIAMENT AND OF THE COUNCIL of 8 May 2003 on the promotion of the use of biofuels or other renewable fuels for transport. *Official Journal of the European Union*.

Fachverband Mineralölindustrie, 2010a. *Austrian fuel consumption statistics*, Wirtschaftskammer Österreich. Available at: http://portal.wko.at/wk/startseite_dst.wk?dstid=308.

Fachverband Mineralölindustrie, 2010b. *Average yearly fuel prices in Austria*, Wirtschaftskammer Österreich. Available at: http://portal.wko.at/wk/startseite_dst.wk?dstid=308.

Fischer, W., 1994. Neue Batteriesysteme. In *Elektrische Straßenfahrzeuge - Technik Entwicklungsstand und Einsatzgebiete*. Kontakt & Studium. Expert Verlag.

Foley, A., Daly, H. & Gallichor, B.O., 2010. Quantifying the Energy & Carbon Emissions Implications of a 10% Electric Vehicles Target. In International Energy Workshop. Stockholm.

Fraunhofer ISI, 2010. Wirtschaftlichkeit von Ladeinfrastrukturkonzepten. *Fraunhofer Elektromobilitätsforschung*. Available at: http://www.fraunhofer-isi-cms.de/elektromobilitaet/Media/forschungsergebnisse/12795511297019-77.22.102.21-6.1.3._Beladungsinfrastruktur.pdf.

Frey, H. et al., 2007. *Pricing - Verkehr nachhaltig steuern*, VCOE.

Fulton, L., Cazzola, P. & Cuenot, F., 2009. IEA Mobility Model (MoMo) and its use in the ETP 2008. *Energy Policy*, 37(10), pp.3758-3768.

GM, 2010. Volt Produktionsmodell. *Chevrolet Webpage*. Available at: http://www.chevrolet.at/entdecken-sie-chevrolet/blick-in-die-zukunft/volt-production-model.html.

Goodwin, P., Dargay, J. & Hanley, M., 2004. Elasticities of Road Traffic and Fuel Consumption with Respect to Price and Income: A Review. *Transport Reviews*, 24(3), pp.275-292.

Greening, L.A. & Bataille, C., 2009. Bottom-up models of energy across the spectrum. In *The International Handbook on the Economics of Energy*. Cheltenham: Edward Elgar Publishing.

Grübler, A., 1998. *Technology and Global Change*, Cambridge.

Haas, R., Ajanovic, A. et al., 2008. *ALTANKRA Project Final Report*, Vienna University of Technology. Available at: http://verkehrstechnologien.at/altankra/_/prog1/subprog25/project453.

Haas, R., Nakicenovic, N. et al., 2008. Towards sustainability of energy systems: A primer on how to apply the concept of energy services to identify necessary trends and policies. *Energy Policy*, 36(11), pp.4012-4021.

Haas, R. & Wirl, F., 1992. Verbraucherseitige Energiesparstrategien - Ein analytischer Überblick. *Zeitschrift für Energiewirtschaft*, 16(1), pp.23-32.

Helmers, E., 2009. *Bitte wenden Sie jetzt - Das Auto der Zukunft*, Wiley-VCH.

Helmolt, R. & Eberle, U., 2007. Fuel cell vehicles: Status 2007. *Journal of Power Sources*, 165(2), pp.833-843.

Herry, M., Sedlacek, N. & Steinbacher, I., 2007. *Verkehr in Zahlen*, Federal Ministry of Transport Innovation and Technology.

Hofmann, P., 2008. *Hybridfahrzeuge*, Vorlesungsskriptum.

Howell, D., 2009. *Progress Report for Energy Storage Research and Development FY2008*, U.S. Department of Energy.

hybridcars.com, 2010. Global Hybrid Car Sales Statistics. Available at: http://www.hybridcars.com/.

IEA, 2008. *Energy Technology Perspectives 2008*, International Energy Agency.

IEA, 2009. Hybrid and Electric Vehicles - The electric drive establishs a market foothold, IEA-Hybrid and Electric Vehicle Implementing Agreement.

IEA, 2007. *IEA Technology Essentials - Fuel Cell*, International Energy Agency IEA.

IEA, 2009a. Key World Energy Statistics 2009, IEA.

IEA, 2009b. *World Energy Outlook 2009*, International Energy Agency IEA.

IER, 2009. *Entwicklungsstand und Perspektiven der Elektromobilität*, Universität Stuttgart - Institut für Energiewirtschaft und Rationelle Energieanwendung Prof. Dr. Ing. A. Voß.

Jaccard, M., 2009. Combining bottom up and top down in energy economy models. In *The International Handbook on the Economics of Energy*. Cheltenhaml: Edward Elgar Publishing.

Johansson, O. & Shipper, L., 1997. Measuring the Long-Run Fuel Demand of Cars. *Transport Economics and Policy*, 31(3).

Jutila, S. & Jutila, J., 1986. Diffusion of Innovation in American Automobile Industry, In Paper prepared for The Advanced Summer Institute in Regional Science. University of Umea, Sweden.

Kalhammer, F.R. et al., 2007. Status and Prospect for Zero Emission Technology - Report of the ARB Independent Expert Panel 2007, State of California Air Ressource Board.

Kapros, P. et al., 2008. *European Energy and Transport - Tends to 2030 - Update 2007*, European Commission Directorate-General for Energy and Transport.

Kloess, M., 2010. The role of plug-in-hybrids as bridging technology towards pure electric cars: An economic assessment. In 25th Electric Vehicle Symposium. Shenzhen.

Kloess, M., 2006. Zukunftsstrategien der Automobilindustrie im Bereich alternativer Antriebe und Kraftstoffe, Vienna University of Technology.

Kloess, M. et al., 2009. ELEKTRA-Project final Report. Available at: http://verkehrstechnologien.at/elek-tra/_/prog46/subprog41/project616.

Kloess, M. & Müller, A., 2011. Simulating the impact of policy, energy prices and technological progress on the passenger car fleet in Austria - Model-based analysis 2010-2050. *Energy Policy*, submitted 2010.

Kloess, M., Müller, A. & Haas, R., 2010. The effects of policy, energy prices and technological learning on the passenger vehicle sector in Austria - A model based analysis. In The International Energy Workshop. Stockholm.

Kloess, M., Weichbold, A. & Könighofer, K., 2009. Technical, Ecological and Economic Assessment of Electrified Powertrain Systems for Passenger Cars in a Dynamic Context (2010 to 2050). In EVS24 International Battery, Hybrid and Fuel Cell Electric Vehicle Symposium. Stavanger.

Köhler, U., 2007. Batterien für Elektro- und Hybridfahrzeuge. In *Hybrid-, Batterie- und Brennstoffzellenfahrzeuge*. Kontakr & Studium. Expert Verlag.

Kojima, K. & Morita, T., 2010. Development of Fuel Cell Hybrid Vehicle in Toyota. In *EVS 25 Proceedings*. World Battery, Hybrid and Fuel Cell Electric Vehicle Symposium & Exhibition EVS 25. Shenzhen.

Kromer, M.A. & Heywood, J.B., 2007. *Electric Powertrains: Opportunities and Challenges in the U.S. Light-Duty Vehicle Fleet*, Sloan Automotive Laboratory Laboratory for Energy and the Environment Massachusetts Institute of Technology.

Lescaroux, F. & Rech, O., 2008. The Impact of Automobile Diffusion on the Income Elasticity of Motor fuel Demand. *The Energy Journal*, 29(1).

Litzlbauer, M., 2009. *Erstellung und Modellierung stochastischer Ladeprofile mobiler Energiespeicher mit MATLAB*, Vienna: Vienna University of Technology - Institute of Powersystems.

Madani, A., 2009. HEV, P-HEV&EV-MARKET-TRENDS 2008-2020 Battery is the key!

Maibach, M. et al., 2008. Handbook on estimation of external costs in the transport sector, CE Delft.

Matheys, J. & Autenboer, W.V., 2005. *SUBAT Sustainable Batteries*,

McDonald, A. & Schrattenholzer, L., 2001. Learning rates for energy technologies. *Energy Policy*, 29(4), pp.255-261.

Meyer, I. & Wessely, S., 2009. Fuel efficiency of the Austrian passenger vehicle fleet-- Analysis of trends in the technological profile and related impacts on CO2 emissions. *Energy Policy*, 37(10), pp.3779-3789.

Miller, J.F., 2010. Analysis of Current and Projected Battery Manufacturing Costs for Electric, Hybrid, and Plug-in Hybrid Electric Vehicles. In The 25th World Battery and Electric Vehicle Symposium EVS 25. Shenzhen.

MIT, 2008. *On the road in 2035*, Laboratory of Energy and the Environment - Massechusetts Institute ot Technology. Available at: http://web.mit.edu/sloan-auto-lab/research/beforeh2/otr2035/.

Mitsubishi Motor Corporation, 2010. i MiEV - Mitsubishi innovative Electric Vehicle. Available at: http://www.mitsubishi-motors.com/special/ev/index.html.

Mock, P. et al., 2009. Electric vehicles – A model based assessment of future market prospects and environmental impacts. In EVS24 International Battery, Hybrid and Fuel Cell Electric Vehicle Symposium. Stavanger.

Nakicenovic, N., 1991. Diffusion of Pervasive Systems: A case of Transport Infrastructures. In *Diffusion of Technologies and Social Behaviour*. Laxemburg: Springer.

Nakicenovic, N., 1987a. Dynamics of Replacement of US Transport Infrastructure. In *Cities and their vital Systems, Infrastructure, Past, Present and Future*. Washington DC: National Academy Press.

Nakicenovic, N., 1987b. Technological Substitution and Long Waves in the USA. In *The Long Wave Debate*. Berlin: Springer.

Nakicenovic, N., 1986. The automobile road to technological change : Diffusion of the automobile as a process of technological substitution. *Technological Forecasting and Social Change*, 29(4), pp.309-340.

Naunin, D., 1994. *Elektrische Straßenfahrzeuge*, Expert Verlag.

Nemry, F. et al., 2008. Environmental Improvement of Passenger Cars (IMPRO-car), EU-JRC, IPTS.

Odyssee, 2010. Energy Efficiency Indicators in Europe -. Available at: http://www.odyssee-indicators.org/.

OICA, 2010. International Organisation of Motor Vehicle Manufacturers - OICA. Available at: http://www.oica.net/category/production-statistics/.

Passier, G. et al., 2007. Status Overview of Hybrid and Electric Vehicle technology (2007), IEA.

Pazdernik, K. et al., 2009. *Emissionstrends 1990-2007*, Wien: Umweltbundesamt.

Peled, E. et al., Parameter analysis of a practical lithium- and sodium-air electric vehicle battery. *Journal of Power Sources*, In Press, Corrected Proof. Available at: http://www.sciencedirect.com/science/article/B6TH1-5166699-7/2/4809d2ae82e61b3b6161e2b441d398de.

Pischinger, F., 2007. Grundlagen der Motorentechnik. In *Vieweg Handbuch Kraftfahrzeugtechnik*. ATZ/MTZ Fachbuch. Wiesbaden: Vieweg.

Pöppl, G. et al., 2009. *Energieinfrastruktur für die Bahn der Zukunft*, Klima & Energiefonds - Energie der Zukunft.

Pötscher, F., 2009. *CO2 Monitoring 2009*, Umwelbundesamt.

Ragwitz, M. et al., 2009. EmployRES - The impact of renewable energy policy on economic growth and employment in the European Union, Kralsruhe.

Rogers, E.M., 2003. *Diffusion of Innovations* 5. ed., New York: Free Press.

Salchenegger, S., 2006. *CO2 Monitoring 2005*, Umweltbundesamt.

Sarnes, J., 1992. Wirtschaftlichkeit von Elektrostraßenfahrzeugen. In *Elektro-Straßenfahrzeuge*. VDI Berichte. VDI Verlag.

Satorius, C. & Zundel, S., 2005. Time Strategies Innovation and Environmental Policy, Edward Elgar.

Schipper, L., Marie-Lilliu, C. & Fulton, L., 2002. Diesel in Europe: Analysis of Characteristics, Usage Patterns, Energy Savings and CO2 Emission Implications. *Journal of Transport Economics and Policy*, 36(2), pp.305-340.

Schneider, J. & Wappel, D., 2009. *Klimaschutzbericht 2009*, Austrian Ministry of Environment - Umweltbundesamt.

Schoots, K., Kramer, G. & van der Zwaan, B., 2010. Technology learning for fuel cells: An assessment of past and potential cost reductions. *Energy Policy*, 38(6), pp.2887-2897.

Schott, B., 2010. *Lithium - Begehrter Rohstoff der Zukunft - eine Verfügbarkeitsanalyse*, Baden-Württemberg: Zentrum für Sonnenenergie- und Wasserstoff-Forschung.

Schwoon, M., 2008. Learning by doing, learning spillovers and the diffusion of fuel cell vehicles. *Simulation Modelling Practice and Theory*, 16(9), pp.1463-1476.

Seiffert, U., 2007a. Fahrzeugphysik. In *Vieweg Handbuch Kraftfahrzeugtechnik*. ATZ/MTZ Sachbuch. Vieweg.

Seiffert, U., 2007b. Produktinnovation, bisherige Fortschritte. In *Vieweg Handbuch Kraftfahrzeugtechnik*. ATZ/MTZ Sachbuch. Vieweg, p. 13.

Sorger, H. et al., 2009. The AVL Pure Range Extender - Layout Development and Results.

Sorrell, S., 2009. The rebound effect: definition and estimation. In *International Handbook on the Economics of Energy*. Cheltenham: Edward Elgar Publishing.

Statistics Austria, 2010a. Austrian Consumer Price Index. Available at: http://www.statistik.at/web_en/.

Statistics Austria, 2010b. Austrian Energy Efficiency Indicators. Available at: http://www.statistik.at/web_de/statistiken/energie_und_umwelt/energie/energieeffizienzi ndikatoren/036570.html.

Statistics Austria, 2010c. Austrian Gross Domestic Product. Available at: http://www.statistik.at/.

Statistics Austria, 2009a. Austrian Household Energy Demand Statistics. Available at: http://www.statistik.at/web_en/.

Statistics Austria, 2009b. Austrian motor vehicle fleet & registration statistics. Available at: http://www.statistik.at/.

Storchmann, K., 2005. Long-Run Gasoline demand for passenger cars: the role of income distribution. *Energy Economics*, 27(1), pp.25-58.

Strock, J.M., 1996. Californias Zero Emission Vehicle (ZEV) Legislation. In *The Future of the Electric Vehicle*. StromDISKUSSION Information zur Elektrizität IZE.

Timm, H. & König, B., 2008. Development Tendencies for Lightweight Design of the Future.

Train, K.E., 2009. *Discrete Choice Models with Simulation* 2. ed., New York: Cambridge University Press.

Tsuchiya, H. & Kobayashi, O., 2004. Mass production cost of PEM fuel cell by learning curve. *International Journal of Hydrogen Energy*, 29(10), pp.985-990.

Turcksin, L. et al., 2008. Market Potential of Clean Vehicles. In EET-2008 European Ele-Drive Conference International Advanced Mobility Forum. Geneva.

US Departement of Energy, 2010. Alternative Fuels & Advances Vehicels Data Center. Available at: http://www.afdc.energy.gov/afdc/data/vehicles.html.

USABC, 2009. USABC - Energy Storage System Goals. Available at: http://www.uscar.org/guest/article_view.php?articles_id=85 [Accessed January 12, 2010].

Van den Brink, R.M.M. & Van Wee, B., 2001. Why has car-fleet specific fuel consumption not shown any decrease since 1990? Quantitative analysis of Dutch passenger car-fleet specific fuel consumption. *Transportation Research Part D: Transport and Environment*, 6(2), pp.75-93.

VCÖ-Forschungsinstitut, 2009. *Potenziale von Elektromobilität*, Wien: Verkehrsclub Österreich VCÖ.

VDA, 2010. Verband der Automobilindustrie - VDA. *VDA Automobilproduktion*. Available at: http://www.vda.de/de/zahlen/jahreszahlen/automobilproduktion/.

van Vliet, O.P. et al., 2010. Techno-economic comparison of series hybrid, plug-in hybrid, fuel cell and regular cars. *Journal of Power Sources*, 195(19), pp.6570-6585.

Voy, C., 1992. Demonstrationsprojekt zur Erprobung von Elektrofahrzeugen der neuesten Generation auf der Insel Rügen. In *Elektro-Straßenfahrzeuge*. VDI Berichte. Düsseldorf: VDI Verlag.

Wagner, U., 1988. Realisierungsaussichten des Elektroautos. *Elelktrische Bahnen*, (Heft 8/1988), p.239.

Wallentowitz, H., Freialdenhoven, A. & Olschewski, I., 2010. *Strategien zur Elektrifizierung des Antriebstranges* 1. ed., Wiesbaden: Vieweg+Teubner.

Wandt, H., 2008. Toyota Prius. In *Hybrid-, Batterie- und Brennstoffzellenfahrzeuge*. Expert-Verlag.

WBCSD, 2004. *Mobility 2030: Meeting the challenges to sustainability*, World Business Council for Sustainable Development.

Weider, M., Metzner, A. & Rammler, S., 2004. Das Brennstoffzellenrennen - Aktivitäten und Strategien bezüglich Brennstoffzellen in der Automobilindustrie, Berlin: Wissenschaftsszentrum Berlin für Sozialforschung.

Werber, M., Fischer, M. & Schwartz, P.V., 2009. Batteries: Lower cost than gasoline? *Energy Policy*, 37(7), pp.2465-2468.

Wietschel, M. & Dollinger, D., 2008. Quo Vadis Elektromobilität? *Energiewirtschaftliche Tagesfragen*, (12), pp.8-15.

Wietschel, M., Hasenauer, U. & de Groot, A., 2006. Development of European hydrogen infrastructure scenarios--CO2 reduction potential and infrastructure investment. *Energy Policy*, 34(11), pp.1284-1298.

Williams, B.D. & Kurani, K.S., 2007. Commercializing light-duty plug-in/plug-out hydrogen-fuel-cell vehicles: "Mobile Electricity" technologies and opportunities. *Journal of Power Sources*, 166(2), pp.549-566.

Winter, R., 2008. Biokraftstoffe im Verkehrssektor in Österreich 2008, Umweltbundesamt.

World Energy Council, 2007. *Transport Technologies and Policy Scenarios to 2050*, World Energy Council.

Wurster, R. & et al., 2002. GM Well-to-Wheel Analysis of Energy Use and Greenhouse gas Emissions of Advanced Fuel/Vehicle Systems - A European Study, L-B-Systemtechnik.

Zachariadis, T., 2005. Assessing policies towards sustainable transport in Europe: an integrated model. *Energy Policy*, 33(12), pp.1509-1525.

Zervas, E. & Lazarou, C., 2008. Influence of European passenger cars weight to exhaust CO2 emissions. *Energy Policy*, 36(1), pp.248-257.

11 Appendix A

11.1 Taxes on fuels in passenger cars in EU member states

Table A-1: Taxation of Vehicles in the EU-27 (Source: Altermotive Coutry Report (Ajanovic et al. 2009))

	VAT	Tax on Acquisition	Tax on Ownership
Austria	20%	Based on fuel consumption Maximum 16% + bonus/malus	kW
Belgium	21%	Based on cc + age	Cylinder Capacity
Bulgaria	20%	68-124€ (depending on age)	kW
Cyprus	15%	Based on cc + CO2	Cylinder Capacity
Czech Republic	19%	None	None
Denmark	25%	105% up to DKK 79,000 180% on the remainder	Fuel Consumption, Weight
Estonia	18%	None	None
Spain	16%	Based on CO2 emissions From 0% (up to 120g/km) to 14.75% (above 200g/km)	Horsepower
Finland	22%	Based on price + CO2 emissions Tax % = 4.88 + (0.122 x CO2) Min. 12.2%, max. 48.8 %	Age, Fuel, Weight
France	19,6%	Based on CO2 emissions From € 200 (161 to 165g/km) to € 2,600 (above 250g/km)	CO2 Emissions
Germany	19%	None	Cylinder capacity, exhaust emissions CO2 emissions
Greece	19%	Based on cc + emissions 5% - 50%	Cylinder Capacity
Hungary	25%	Based on emissions	Weight
Ireland	21,5%	Based on CO2 emissions max. 36%	CO2 Emissions
Italy	20%	IPT + PRA + MCTC	kW, Exhaust Emissions
Lithuania	19%	None	None
Luxembourg	15%	None	CO2 Emissions
Latvia	21%	€ 373	Weight
Malta	18%	Based on price, CO2 emissions, vehicle length	Cylinder Capacity
Netherlands	19%	Based on price + CO2 emissions 40% - € 1, 394(petrol) 40% + € 290 (diesel)	Weight, porvince
Poland	22%	Based on cc 3.1% - 18.6%	None
Portugal	20%	Based on cc + CO2 emissions	Cylinder Capacity, CO2 Emissions
Romania	19%	Based on cc + emissions + CO2	Cylinder Capacity
Sweden	25%	None	CO2 Emissions, Weight
Slovenia	20%	Based on price 1% –13%	None
Slovakia	19%	None	None
UK	15%	None	CO2 Emissions, Cylinder Capacity

Table A-2: Fuel Taxes in EU-27 countries (Data Source: (European Commission 2010))

	Unleaded Gasoline (€/1000l)	Diesel (€/1000l)
Austria	442	347
Belgium	592	318
Bulgaria	350	307
Cyprus	299	245
Czech Republic	483	406
Denmark	561	382
Estonia	359	330
Spain	360	330
Finland	627	364
France	607	428
Germany	655	470
Greece	359	302
Hungary	448	368
Ireland	509	368
Italy	564	423
Lithuania	434	330
Luxembourg	462	302
Latvia	379	330
Malta	459	352
Netherlands	701	413
Poland	488	339
Portugal	583	364
Romania	336	284
Sweden	468	446
Slovenia	403	383
Slovakia	515	481
UK	661	661

11.2 Physical background of fuel consumption of cars

For a better understanding of the fuel efficiency problem of motor vehicles a short view on the physical background of fuel consumption will be given:

The power a car requires for driving can be defined as follows:

$$P_{total} = \frac{P_W}{\eta_D} + P_{aux} \quad [kW] \tag{A-1}$$

$$P_W = F_W \cdot v \quad [kW] \tag{A-2}$$

P_{total} ... total energy required by the vehicle [kW]
P_W ... power required to overcome all driving resistances [kW]
P_{aux} ... power required for auxiliaries [kW]
η_D ... drivetrain efficiency (motor & drivetrain) [%]
v ... vehicle speed / speed of air flow [km h^{-1}]

Where P_w is the power needed to overcome the driving resistances F_w, $\eta_{drivetrain}$ is the efficiency of the drivetrain including motor and transmission and P_{aux} is the power required for the auxiliaries.

$$F_W = F_{RO} + F_L + F_{ST} + F_B \tag{A-3}$$

F_W ... total driving resistance
F_{RO} ... rolling resistance
F_L ... aerodynamic drag
F_{ST} ... climbing resistance
F_B ... acceleration resistance

When the car is driving at constant speed the driving resistance F_w is determined by the rolling resistance F_{RO}, the aerodynamic drag F_L and the climbing resistance F_{ST}. The rolling resistance of the tyres F_{RO} depends on the vehicle mass, and the rolling resistance coefficient, which is depending on the quality of tyres and the state of the road:

$$F_{RO} = f \cdot m \cdot g \tag{A-4}$$

F_{RO} ... rolling resistance
f ... rolling resistance coefficient
m ... vehicle mass
g ... gravitational acceleration

The aerodynamic drag is defined by the vehicle's driving speed (respectively by the speed of the air flow around the vehicle), the windage area and the aerodynamic coefficient of the car:

$$F_L = c_W \cdot A \cdot \rho \cdot \frac{v^2}{2} \tag{A-5}$$

F_L ... aerodynamic drag
c_W ... aerodynamic drag coefficient
ρ ... air density
v ... vehicle speed / speed of air flow
A ... windage area

The climbing resistance is often neglected since test cycles (e.g. NEDC) have no climbing sections. As indicated by equation (A-6) climbing resistance is determined by the angle of elevation and the vehicle mass.

$$F_{St} = m \cdot g \cdot \sin\beta \tag{A-6}$$

F_{ST} ... climbing resistance
m ... vehicle mass
g ... gravitational acceleration
β ... angle of elevation

When the car is accelerating total resistance is affected by a further term, the acceleration resistance. The acceleration resistance that is dependent on the vehicle mass and the mass moments of inertia of rotating elements in the car.

$$F_B = m_{red} \cdot dv/dt \tag{A-7}$$

F_B ... acceleration resistance
m_{red} ... dynamic mass considering mass moments of inertia of rotating elements

Summing up the resistances the fuel consumption can be expressed in one equation (Seiffert 2007b):

$$B_e = \frac{\int b_e \cdot \frac{1}{\eta_D} \left[\left(m \cdot f \cdot g \cdot \cos\beta + \frac{\rho}{2} \cdot c_W \cdot A \cdot v^2 \right) + m \cdot (a + \sin\beta) + B_r \right] \cdot v \cdot dt}{\int v \cdot dt} \tag{A-8}$$

B_e ... fuel consumption of the vehicle
b_e ... specific energy demand of the motor
η_D ... drivetrain efficiency
B_r ... breaking resistance

Table A-3: Efficiency of energy conversion steps in the powertrain systems

		ICE-CD		ICE-HEV		ICE-SHEV		BEV		FC		Source
Primary Energy Source		Crude Oil	Crude Oil	Crude Oil	Crude Oil	Crude Oil	Crude Oil	Natural Gas	RES	Natural Gas	RES	
Final Energy Carrier		Gasoline	Diesel	Gasoline	Diesel	Gasoline	Diesel	NG	RES	NG	RES-H2	
WTT												
Fuel Production	Gasoline	81%		81%		81%						Joanneum Research 2009
	Diesel		90%		90%		90%					Joanneum Research 2009
	Electricity							43%				Joanneum Research 2009
	Hydrogen									68%		Van Mierlo et al. 2006
Refuelling	Gasoline/Diesel	100%	100%	100%	100%	100%	100%					Joanneum Research 2009
	Battery Charging					90%	90%	90%	90%			Matheys et al 2005
	Hydrogen Compression									75%	75%	Van Mierlo et al. 2006
WTT-Efficiency		81%	90%	81%	90%	81%	90%	28.6%	74%	57.4%	57.4%	
TTW												
Generator ICE	Gasoline					36%						Pischinger et al. 2008
	Diesel						43%					Pischinger et al. 2008
Electric Generator						90%	90%					Wallentowitz 2010
Control AC/DC						97%	97%					Campanari et al. 2009
H2-Fuel Cell										70%	70%	AVL 2009
Battery Charge						90%	90%	90%	90%			Matheys et al 2004
Battery Discharge						90%	90%	90%	90%			Matheys et al 2005
Control DC/AC						97%	97%	97%	97%	97%	97%	Campanari et al. 2009
Electric Motor						90%	90%	90%	90%	90%	90%	Wallentowitz 2010
ICE (NEDC)	Gasoline	22.5%		28.5%								Pischinger et al. 2008
	Diesel		25.2%		32.9%							Pischinger et al. 2008
Drive Train Losses (NEDC)	DI-Gasoline CD	90%										Wallentowitz 2010
	DI-Gasoline HEV (full)			94%		94%						Wallentowitz 2010
	DI-Diesel CD		90%									Wallentowitz 2010
	DI-Diesel HEV (full)				94%		94%					Wallentowitz 2010
	HEV							94%	94%	94%	94%	Wallentowitz 2010
TTW-Efficiency	CD	22.7%	25.7%									
	HEV			32.7%	34.8%	26.8%	32.0%	74%	74%	57.4%	57.4%	
WTW-Efficiency	CD	18.4%	22.7%									
	HEV			23.1%	29.4%	21.7%	28.8%	28.6%	66.5%	29.3%	43.1%	

194

11.3 Detailed specifications of analyzed cars

Table A-4: detailed specifications of compact, middle class and upper class cars

	curb weight [kg]	propulsion system engine/fuel cell [kW]	electric motor [kW]	electric generator [kW]	battery (traction) nominal capacity [kWh]	DOD_min	DOD_max	ΔDOD (usable cap.) [kWh]	mass [kg]	fuel consumption /100km (TTW) electricity [kWh]	fuels [l; kg]	driving range electric [km]	total [km]
Compact Class													
Conventional Drive - gasoline	955	50									6,0		500
Conventional Drive - diesel	989	50									4,7		500
Conventional Drive - CNG	1016	50									4,2		500
Micro Hybrid - gasoline	967	50									5,4		500
Micro Hybrid - diesel	1002	50									4,4		500
Micro Hybrid - CNG	1028	50									3,8		500
BEV 75km	1037		50			80%	20%		240	19,7		73	73
Middle Class													
Conventional Drive - gasoline	1470	75									7,5		700
Conventional Drive - diesel	1522	76									6,0		700
Conventional Drive - CNG	1533	77									5,2		700
Micro Hybrid - gasoline	1495	78									6,9		700
Micro Hybrid - diesel	1547	79									5,7		700
Micro Hybrid - CNG	1558	80									4,8		700
Mild HEV - parallel	1535	65	20						20		6,4		700
Mild HEV - diesel	138	65	20		1				20		5,2		700
Mild HEV - CNG	1598	65	20		1				20		4,4		700
Full HEV - power split - gasoline	1593	50	20		2				30		5,9		700
Full HEV - power split - diesel	1628	50	20		2				30		5,1		700
Full HEV - power split - CNG	1656	50	20		2				30		4,2		700
PHEV power split - 40km - gasoline	1723	50	50	20	16	80%	20%	9,6	160	22,2	5,9	43	700
PHEV series - 40km - gasoline	1608	50	40	25	16	80%	20%	9,6	160	22,2	5,5	43	700
BEV 65km + REX - gasoline	1555	30	75	25	24	80%	20%	14,4	240	22,2	5,5	65	700
BEV 130km	1692		75		48	80%	20%	28,8	480	22,2		130	130
FC PHEV 40km - H2	1784	40	75		16	80%	20%	9,6	160	22,2	0,9	43	700
FCV	1860	80	75		2				30		0,9		500
Upper Class													
Conventional Drive - gasoline	2068	120									9,5		700
Conventional Drive - diesel	2151	120									7,5		700
Conventional Drive - CNG	2138	120									6,7		700
Micro Hybrid - gasoline	2093	120									8,9		700
Micro Hybrid - diesel	2176	120									7,2		700
Micro Hybrid - CNG	2163	120									6,2		700
Mild HEV - parallel	2123	100	50								8,2		700
Mild HEV - diesel	2193	100	50								6,7		700
Mild HEV - CNG	2159	100	50								5,8		700
Full HEV - power split - gasoline	2141	75	75								7,7		700
Full HEV - power split - diesel	2193	75	75								6,5		700
Full HEV - power split - CNG	2211	75	75								5,4		700
PHEV power split - 50km - gasoline	2406	75	120	65	24	80%	20%	14,4	240	28,3	7,7	51	700

195

11.4 Component costs of propulsion systems

Table A-5: Cost of Internal Combustion Engines in the Vehcile Cost Model (Data Source: (EUCAR et al. 2006))

Internal Combustion Engine		120kW			100kW			75kW			65kW			50kW			40kW		
		Gasoline	Diesel	CNG	Gasoline	Diesel	CNG	Gasoline	Diesel	CNG	Gasoline	Diesel	CNG	Gasoline	Diesel	CNG	Gasoline	Diesel	CNG
Power	[kW]	120	120	120	100	100	100	75	75	75	65	65	65	50	50	50	40	40	40
Weight	[kg]	418	501	418	348	418	348	261	313	261	226	272	226	174	209	174	139	167	139
Engine+Transmission	30 [€/kW]	3600	3600	3600	3000	3000	3000	2250	2250	2250	1950	1950	1950	1500	1500	1500	1200	1200	1200
DICI	1500 [€]		1500			1500			1500			1500			1500			1500	
DISI	500 [€]	500			500			500			500			500			500		
Turbo	180 [€]	180	180	180	180	180	180	180	180	180	180	180	180	180	180	180	180	180	180
Friction Improvement	60 [€]	60	60	60	60	60	60	60	60	60	60	60	60	60	60	60	60	60	60
20% Downsizing SI	220 [€]	220			220			220			220			220			220		
Double Injection System bi	700 [€]			700			700			700			700			700			700
EURO IV SI	300 [€]	300			300			300			300			300			300		
EURO IV Diesel	300 [€]																		
EURO IV Diesel + Dpf	700 [€]		700			700			700			700			700			700	
Credit for 3-way catayst	-430 [€]																		
		4860	6040	4840	4260	5440	4240	3510	4690	3490	3210	4390	3190	2760	3940	2740	2460	3640	2440

Table A-6: cost of electric machines (Data Source: (EUCAR et al. 2006))

Electric Machines							
	Power	[kW]	20	40	50	75	
	Weight	[kg]	30	40	50	60	
							Source
Motor		8 [€/kW]	160	320	400	600	EUCAR-CONCAWE-JRC-WTW-2007
Controller		19 [€/kW]	380	760	950	1425	EUCAR-CONCAWE-JRC-WTW-2007
Motor+Controller		27 [€/kW]	540	1080	1350	2025	EUCAR-CONCAWE-JRC-WTW-2007

Table A-7: cost of electric drivetrain adaptation and other components (Data Sources: (EUCAR et al. 2006); (Williams & Kurani 2007))

Hybrid - Powertrain&Vehicle components upgrade

Full Hybrid	2630 [€]	EUCAR-CONCAWE-JRC-2007
Mild Hybrid	1315 [€]	own estimation
Credt for Stanadrad Alternator + Starter	-300 [€]	EUCAR-CONCAWE-JRC-2007
Start/Stop system - Gasoline	200 [€]	EUCAR-CONCAWE-JRC-2007
Start/Stop system - Diesel	300 [€]	EUCAR-CONCAWE-JRC-2007
On-Vehicle Charging System:	690 [$]	Williams & Kurani 2007

Table A-8: cost of tank systems (Data Sources: (EUCAR et al. 2006) (Helmolt & Eberle 2007))

Liquid Fuel Tank					Source	
Capacity	[l]	30	50	70		
Weight (empty)	[kg]	4	7	10	EUCAR-CONCAWE-JRC-2007	
Cost	[€]	100	125	150		

Pressure Vassel 20MPa			Source
Capacity	[kg]	7.5	
Weight (empty)	[kg]	25	EUCAR-CONCAWE-JRC-2007
Cost	[€]	460	

Pressure Vessel 70MPa			Source
Capacity	[kg]	6	GM 2007
Weight (empty)	[kg]	125	
Cost	[€/kg]	575	EUCAR-CONCAWE-JRC-2007
	[€]	3450	EUCAR-CONCAWE-JRC-2007
	[$]	3600	GM 2007

11.5 Economic Assessment

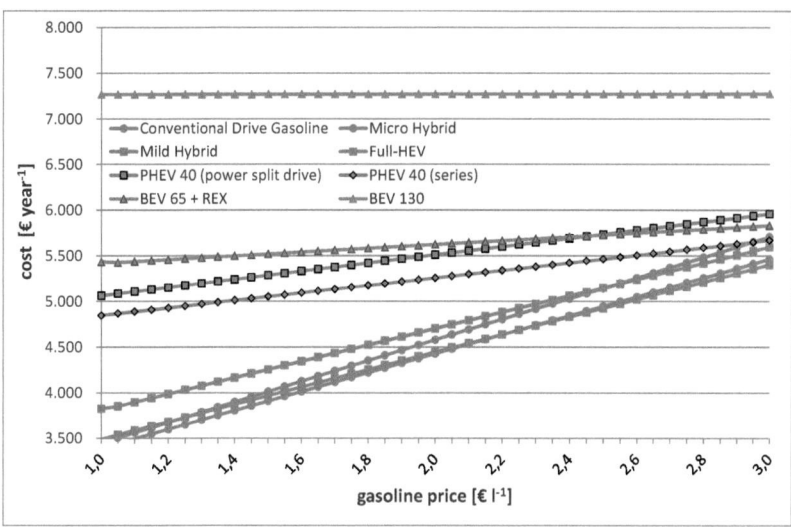

Figure A-1: Gasoline Price Sensitivity of propulsions Systems at technology cost status 2010 (spec. battery cost = 700 € kWh^{-1}; 15 000 km year^{-1})

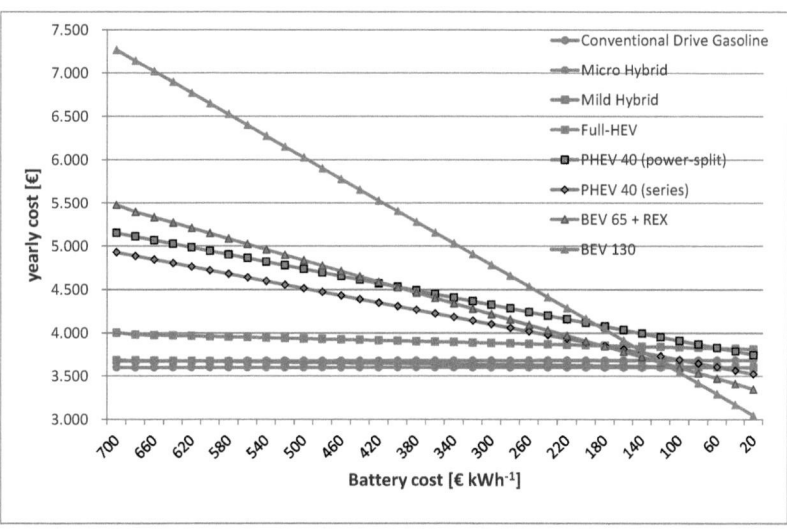

Figure A-2: Battery Cost Sensitivity of Propulsion Systems at a gasoline price of 1.2 € litre^{-1} (15 000 km year^{-1})

11.6 Net Investment Costs – Compact Class Cars

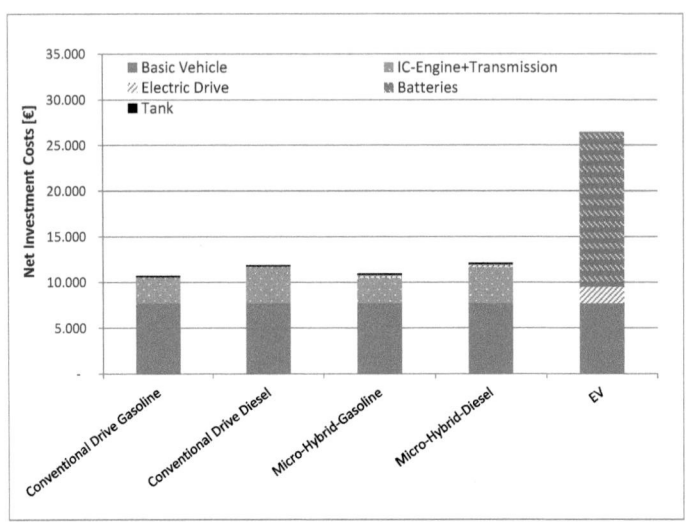

Figure A-3: Net Investment Cost of Powertrain Systems in 2010 (compact class cars)

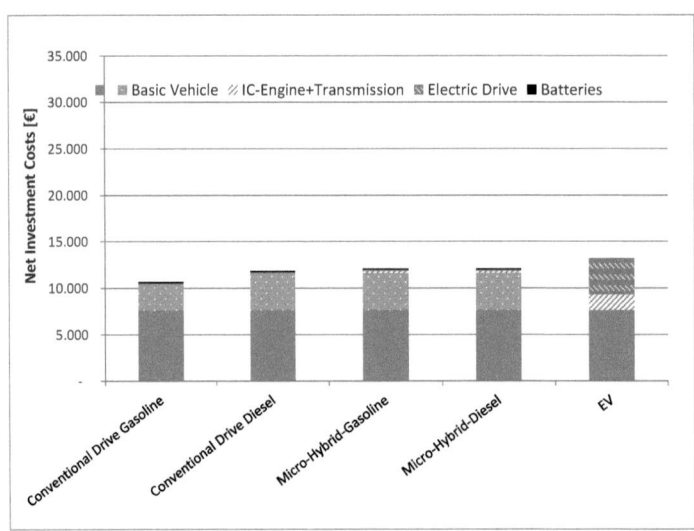

Figure A-4: Net Investment Cost of Powertrain Systems in 2030 (compact class cars)

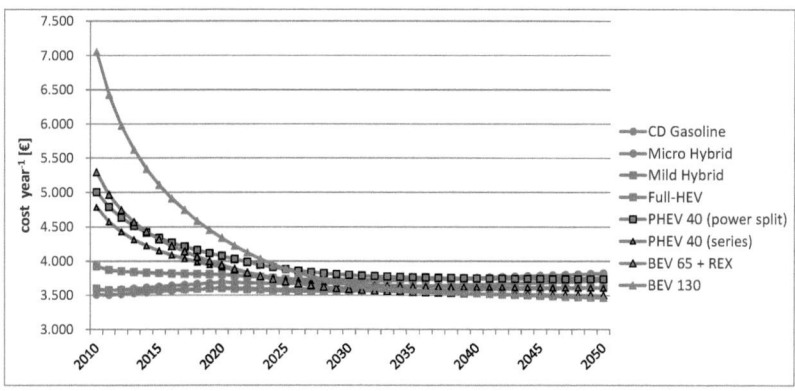

Figure A-5: estimated development of yearly costs of propulsion systems in the middle class 2010 – 2050 in the "Low-Price-Scenario" (15 000 km year^{-1})

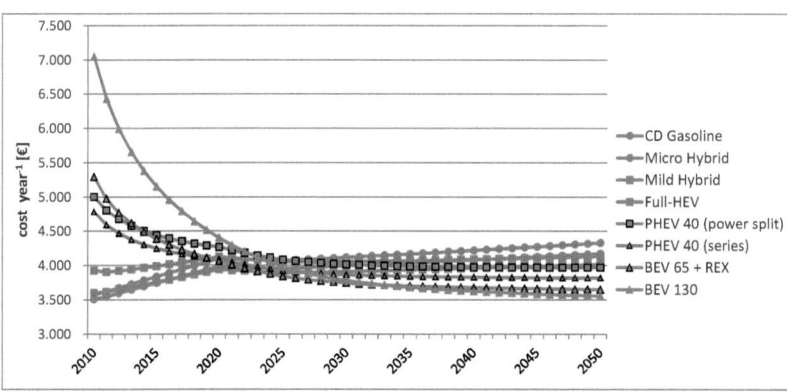

Figure A-6: estimated development of yearly costs of propulsion systems in the middle class 2010 – 2050 in the "High-Price-Scenario" (15 000 km year^{-1})

12 Appendix B

12.1 Model calibration

12.1.1 Market shares of technologies

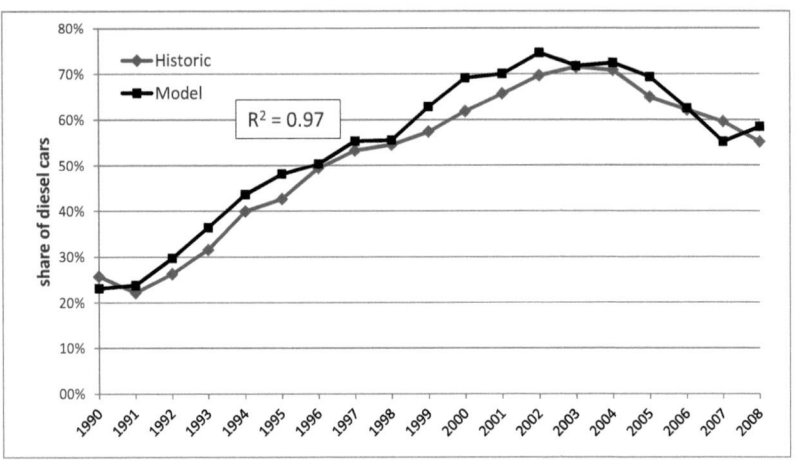

Figure B-1: Share of Diesel cars in Austrian passenger car sales: historic vs. model

The time frame 1990-2009 was used to determine the parameters of the applied top down model of the passenger car fleet. Figure B-2 gives the comparison of the historic development of the passenger car stock (source: (Statistics Austria 2009b)) and the development determined by the model approach in the time frame 1990-2009.

12.1.2 Fleet development

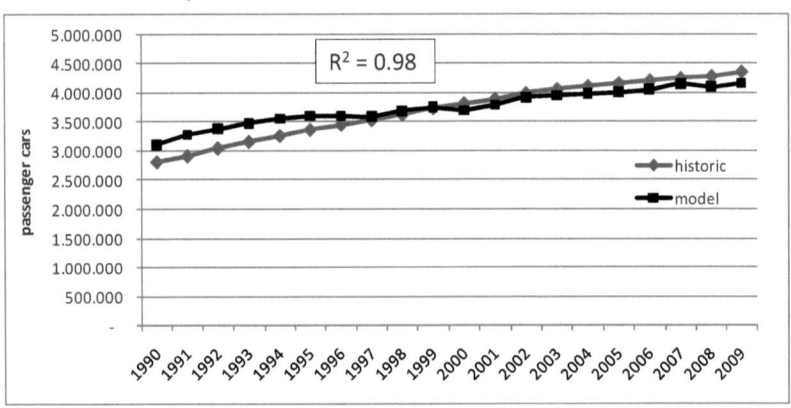

Figure B-2: historic and modelled fleet development

12.1.3 Car characteristics

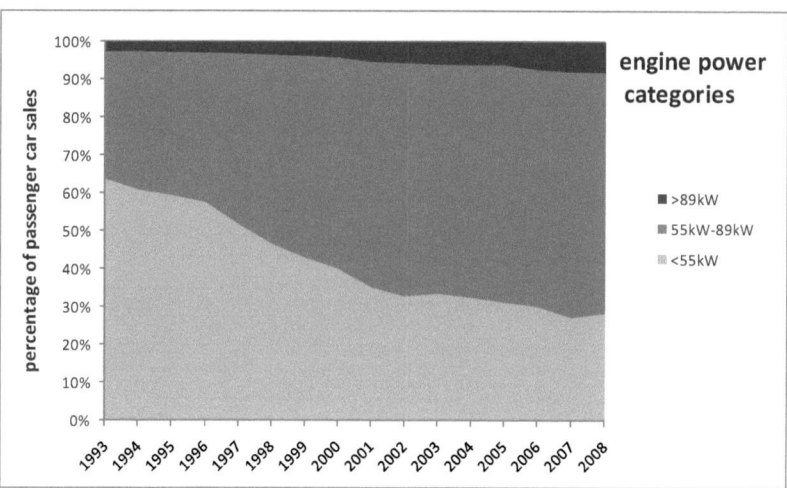

Figure B-3: engine power of new cars in Austria 1993-2008 (data source: (Statistics Austria 2009b))

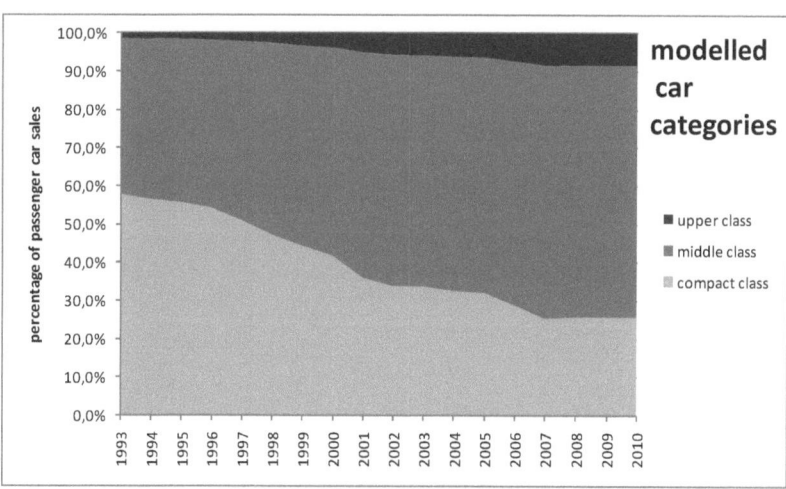

Figure B-4: distribution of vehicle classes in the model 1990-2010

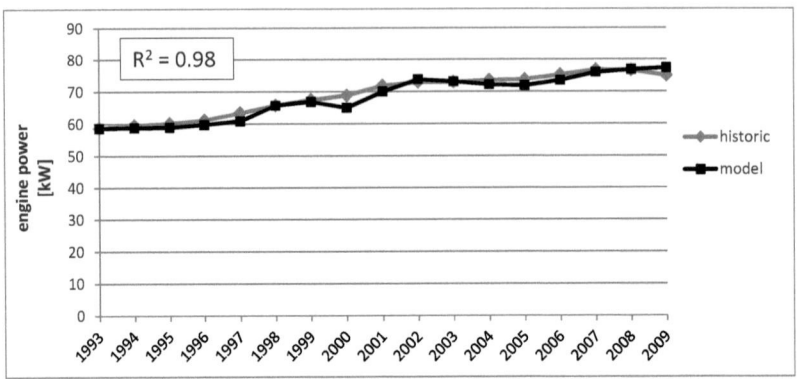

Figure B-5   historic and modelled development of average engine power of passenger cars registered in Austria (data source of historic development: (Statistics Austria 2009b))

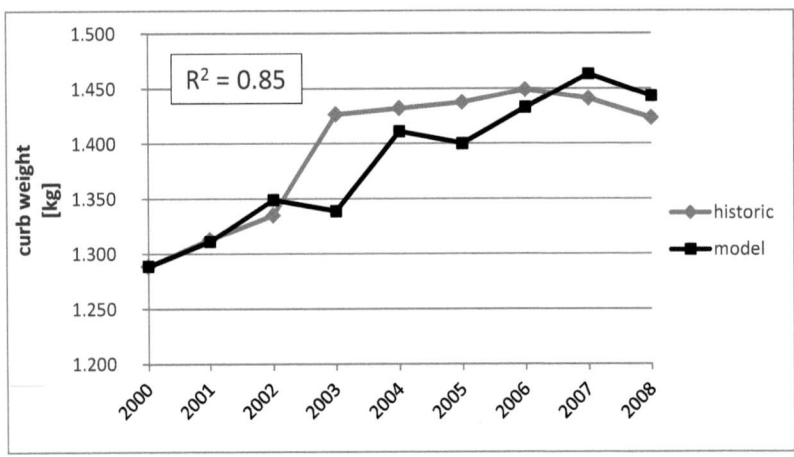

Figure B-6: historic and modelled development of average curb weight of passenger cars sold in Austria (data source of historic development: Uweltbundesamt (Pötscher 2009))

Table B-1: Income and Fuel Price in Austria 1980-2009 (data sources: (Statistics Austria 2010c) (Fachverband Mineralölindustrie 2010b))

	GDP nominal	GDP INDEX 2000	GDP-real 2000	GDP growth	fuels amounts gasoline [t]	diesel total [t]	cars [t]	trucks [t]	prices (nominal) gasoline normal [€/l]	super [€/l]	super plus [€/l]	average [€/l]	diesel [€/l]	prices (real) gasoline real [€/l]	diesel real [€/l]	weighted fuel price index [€/l]
1980	76.60	62.13	123.27		2,436	1,503	94	1,409	0.63		0.65	0.63	0.596	1.02	0.96	1.01
1981	81.60	62.04	131.52	6.7%	2,408	1,446	96	1,350	0.70		0.79	0.75	0.698	1.20	1.13	1.20
1982	87.63	63.25	138.54	5.3%	2,387	1,490	101	1,389	0.76		0.80	0.78	0.698	1.23	1.10	1.23
1983	93.33	65.19	143.17	3.3%	2,466	1,493	109	1,384	0.74		0.80	0.77	0.727	1.18	1.12	1.18
1984	98.01	65.18	150.38	5.0%	2,450	1,425	118	1,307	0.79		0.82	0.80	0.749	1.23	1.15	1.23
1985	103.42	66.84	154.72	2.9%	2,405	1,522	142	1,380	0.80	0.84	0.84	0.83	0.763	1.24	1.14	1.23
1986	108.96	68.32	159.49	3.1%	2,453	1,612	179	1,433	0.67	0.72	0.74	0.71	0.654	1.04	0.96	1.03
1987	113.09	69.37	163.02	2.2%	2,498	1,602	230	1,372	0.60	0.65	0.68	0.64	0.581	0.93	0.84	0.92
1988	118.58	71.77	165.22	1.3%	2,558	1,813	295	1,518	0.58	0.62	0.64	0.61	0.56	0.85	0.78	0.85
1989	126.84	74.31	170.69	3.3%	2,594	1,931	362	1,569	0.63	0.66	0.70	0.66	0.596	0.89	0.80	0.88
1990	135.21	77.73	175.24	2.7%	2,552	2,084	434	1,650	0.68	0.71	0.74	0.71	0.64	0.91	0.82	0.90
1991	146.08	80.52	181.41	3.5%	2,794	2,315	513	1,802	0.64	0.67	0.70	0.67	0.596	0.83	0.74	0.82
1992	154.21	82.43	187.08	3.1%	2,675	2,439	546	1,893	0.68	0.70	0.73	0.70	0.56	0.85	0.68	0.82
1993	159.16	82.70	192.45	2.9%	2,567	2,680	588	2,092	0.68	0.70	0.73	0.70	0.545	0.85	0.66	0.81
1994	167.01	84.90	196.71	2.2%	2,483	3,004	656	2,348	0.71	0.73	0.76	0.74	0.552	0.87	0.65	0.82
1995	174.61	86.52	201.81	2.6%	2,394	2,854	716	2,138	0.78	0.80	0.84	0.81	0.581	0.93	0.67	0.87
1996	180.15	88.79	202.90	0.5%	2,204	3,056	765	2,291	0.82	0.84	0.89	0.85	0.654	0.96	0.74	0.90
1997	183.48	90.42	202.91	0.0%	2,092	3,280	829	2,451	0.85	0.87	0.93	0.89	0.669	0.98	0.74	0.91
1998	190.85	93.64	203.81	0.4%	2,130	3,545	951	2,594	0.80	0.82	0.86	0.83	0.61	0.88	0.65	0.81
1999	197.98	96.75	204.62	0.4%	2,047	3,892	1,029	2,863	0.78	0.79	0.84	0.80	0.6251	0.83	0.65	0.77
2000	207.53	100.00	207.53	1.4%	1,980	4,262	1,133	3,129	0.91	0.93	0.99	0.94	0.778	0.94	0.78	0.88
2001	212.50	100.83	210.75	1.6%	1,998	4,675	1,298	3,377	0.88	0.90	0.97	0.92	0.755	0.91	0.75	0.85
2002	218.85	101.70	215.20	2.1%	2,142	5,175	1,587	3,588	0.83	0.84	0.93	0.86	0.718	0.85	0.71	0.79
2003	223.30	102.93	216.95	0.8%	2,223	5,742	1,851	3,891	0.85	0.86	0.93	0.88	0.729	0.85	0.71	0.79
2004	232.78	105.31	221.05	1.9%	2,133	5,936	1,981	3,955	0.91	0.93	0.99	0.94	0.802	0.90	0.76	0.83
2005	243.58	107.46	226.68	2.5%	2,073	6,264	2,098	4,166	1.00	1.02	1.08	1.04	0.941	0.96	0.88	0.92
2006	256.16	111.01	230.76	1.8%	1,992	6,174	2,409	3,765	1.07	1.09	1.15	1.10	1.008	0.99	0.91	0.95
2007	270.78	114.80	235.87	2.2%	1,966	6,321	2,473	3,848	1.10	1.11	1.18	1.13	1.03	0.99	0.90	0.94
2008	281.87	118.30	238.26	1.0%	1,835	6,148	2,352	3,796	1.21	1.22	1.29	1.24	1.245	1.05	1.05	1.05
2009	277.07	118.90	233.03	-2.2%	1,842	5,952	2,402	3,550	1.05	1.06	1.15	1.09	0.997	0.91	0.84	0.87

203

Table B-2: passenger car fleet and sales in Austria 1980-2009 (data source: (Statistics Austria 2009b))

	passenger car fleet				passenger car sales					
	gasoline	diesel	electric	aggregate	gasoline	diesel	CNG	hybrid	electric	aggregate
1980	2,060,055	73,866	16	2,133,937	220,249	7,299				227,548
1981	2,116,053	78,376	21	2,194,450	190,346	8,313				198,659
1982	2,153,384	84,519	14	2,237,917	190,193	10,962				201,155
1983	2,196,160	90,103	15	2,286,278	244,298	12,378				256,676
1984	2,235,074	100,009	15	2,335,098	201,095	14,545				215,640
1985	2,267,830	123,875	18	2,391,723	209,433	33,237				242,670
1986	2,307,424	156,120	18	2,463,562	219,799	42,376				262,175
1987	2,333,611	198,609	18	2,532,238	190,442	52,779				243,221
1988	2,370,496	253,459	20	2,623,975	190,743	62,329				253,072
1989	2,419,646	312,937	23	2,732,606	208,140	67,925				276,065
1990	2,430,301	382,646	30	2,812,977	214,438	74,197				288,635
1991	2,488,808	423,499	55	2,912,362	236,478	67,245				303,723
1992	2,561,708	483,708	88	3,045,504	236,135	83,959				320,094
1993	2,604,765	552,676	121	3,157,562	195,189	89,973			20	285,162
1994	2,618,845	640,335	131	3,259,311	164,435	109,228			18	273,663
1995	2,633,610	728,995	137	3,362,742	160,419	119,191			13	279,610
1996	2,611,169	838,859	149	3,450,177	155,825	151,846			20	307,671
1997	2,584,583	947,785	163	3,532,531	128,498	146,503			18	275,001
1998	2,565,699	1,060,758	169	3,626,626	134,568	161,297			12	295,865
1999	2,549,804	1,187,137	166	3,737,107	133,954	180,228			3	314,182
2000	2,493,556	1,321,156	156	3,814,868	118,146	191,281			3	309,427
2001	2,428,945	1,460,902	153	3,890,000	100,847	192,681			5	293,528
2002	2,364,743	1,622,350	148	3,987,241	84,938	194,555		16	1	279,493
2003	2,289,547	1,764,760	135	4,054,442	85,616	214,505		8	0	300,121
2004	2,209,315	1,899,814	128	4,109,257	91,037	220,255		133	1	311,292
2005	2,145,831	2,010,912	127	4,156,870	108,007	199,908		260	0	307,915
2006	1,983,337	2,220,804	127	4,204,268	116,830	191,766	112	585		308,596
2007	1,960,380	2,283,302	131	4,243,813	120,466	176,822	247	765	0	297,288
2008	1,957,751	2,323,016	146	4,280,913	131,616	160,459	885	735	2	292,075
2009	1,972,352	2,381,906	223	4,354,481	170,847	146,962	500	1055	39	317,809

12.2 Specific Service costs

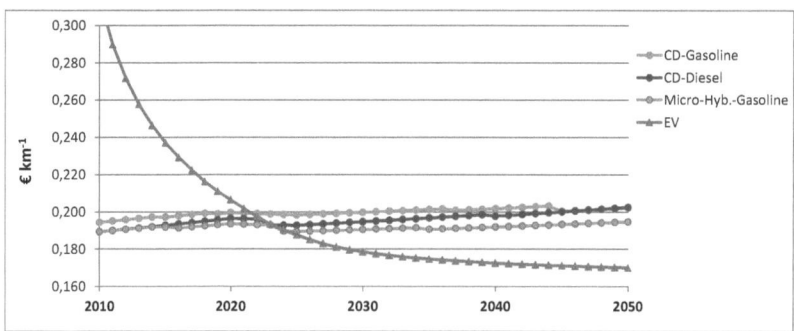

Figure B-7: Specific service cost of compact class cars – BAU & Low Price-Scenario (15 000 km year $^{-1}$)

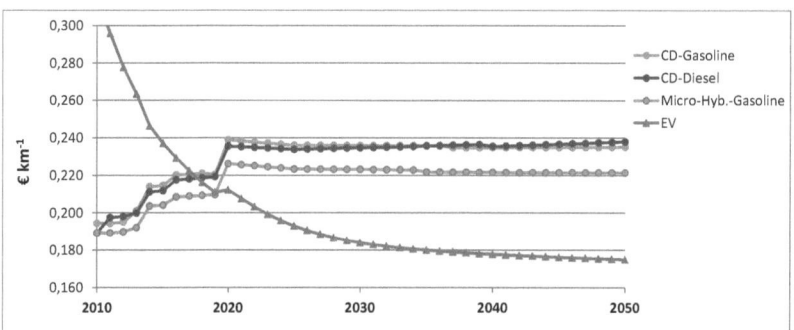

Figure B-8: Specific service cost of compact class cars – Policy & Low Price-Scenario (15 000 km year $^{-1}$)

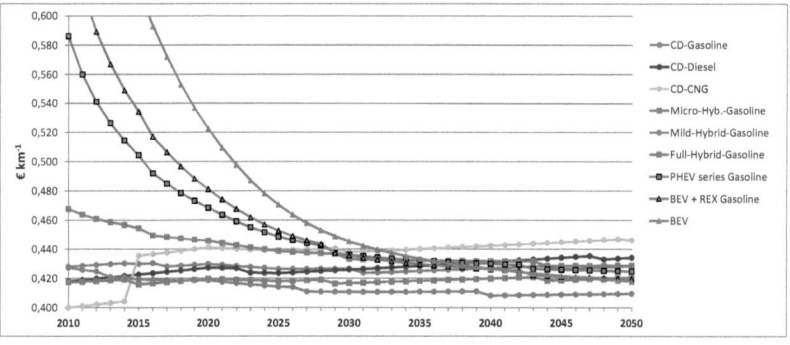

Figure B-9: Specific service cost of middle class cars – BAU & Low Price-Scenario (10 000 km year $^{-1}$)

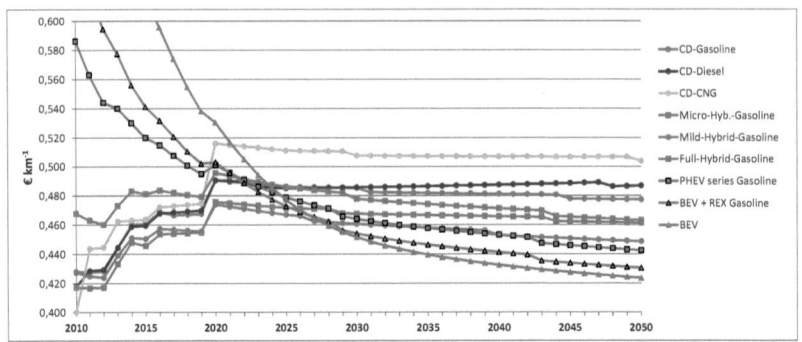

Figure B-10: Specific service cost of middle class cars – Policy & Low Price-Scenario (10 000 km year $^{-1}$)

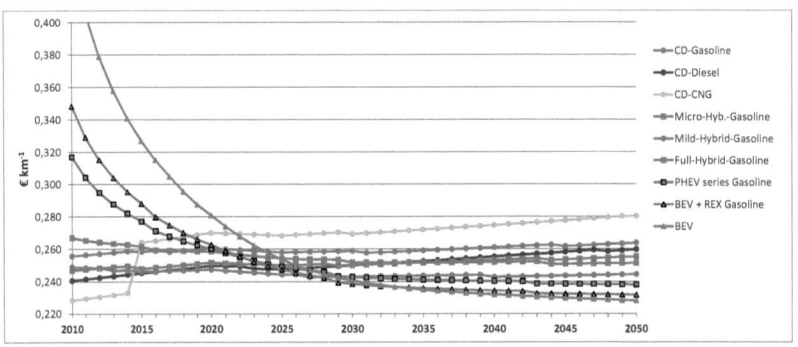

Figure B-11: Specific service cost of middle class cars – BAU & Low Price-Scenario (20 000 km year $^{-1}$)

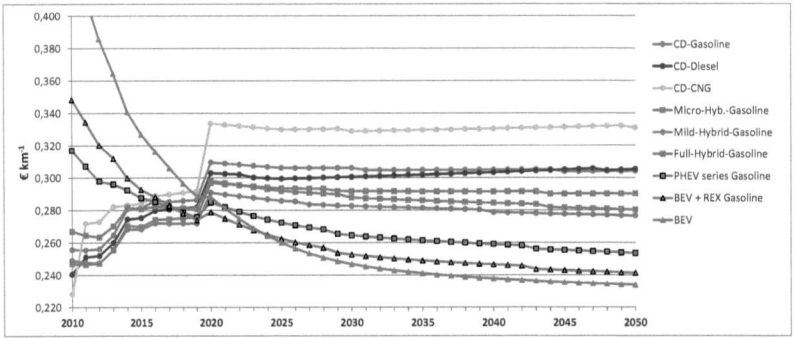

Figure B-12: Specific service cost of middle class cars – Policy & Low Price-Scenario (20 000 km year $^{-1}$)

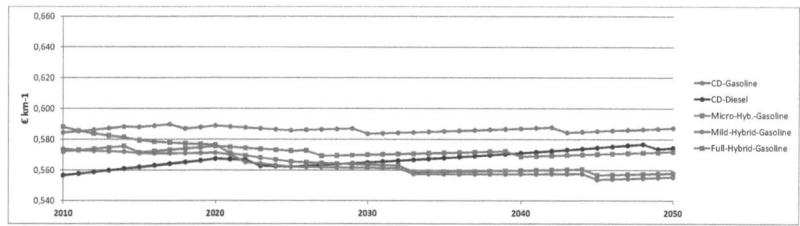

Figure B-13: Specific service cost of upper class cars – BAU & Low Price-Scenario (15 000 km year^{-1})

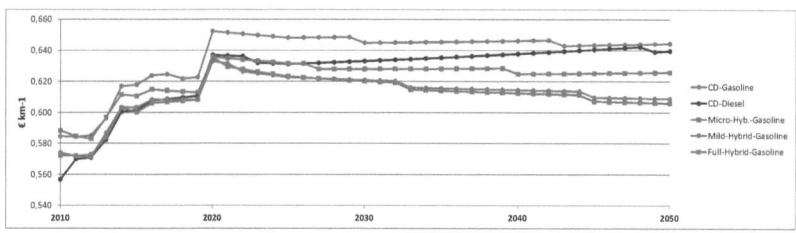

Figure B-14: Specific service cost of upper class cars – Policy & Low Price-Scenario (15 000 km year^{-1})

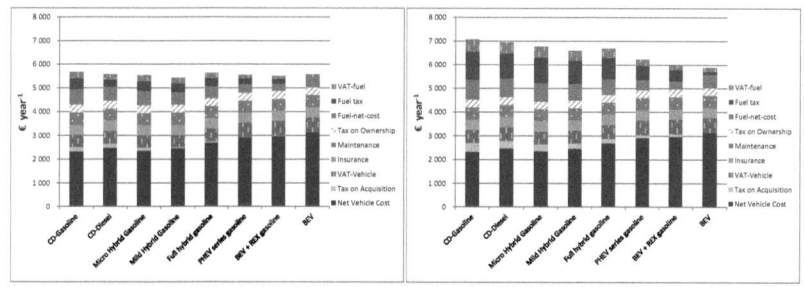

Figure B-15: Total yearly cost of middle class cars with different propulsion systems in the BAU and in the Policy scenario in 2030 (20 000 km year^{-1})

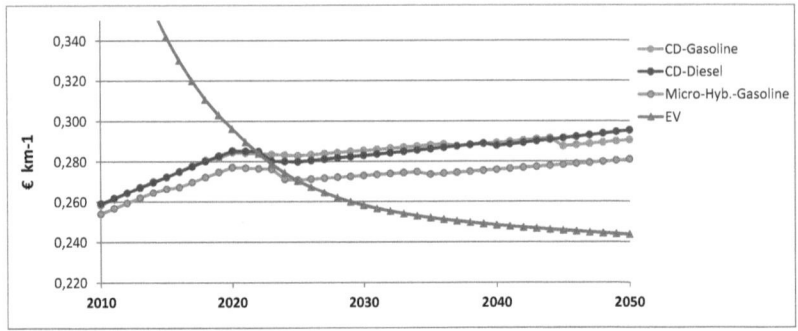

Figure B-16: Specific service cost of compact class cars – BAU & High Price - Scenario (10 000 km year $^{-1}$)

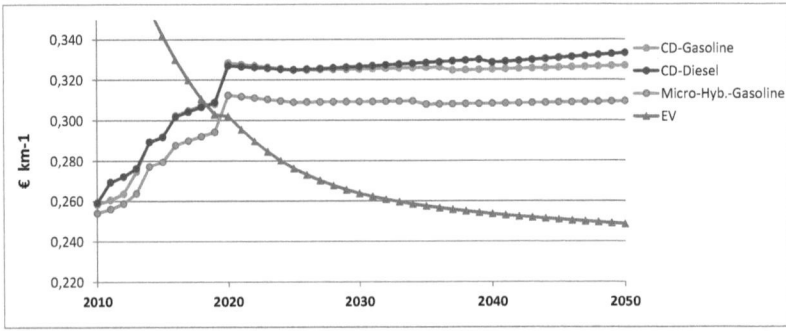

Figure B-17: Specific service cost of compact class cars – Policy & High Price - Scenario (10 000 km year $^{-1}$)

12.3 Market and fleet penetration

12.3.1 "Business as usual"-Policy & moderate fossil fuel price increase (BAU + Low Price-Scenario)

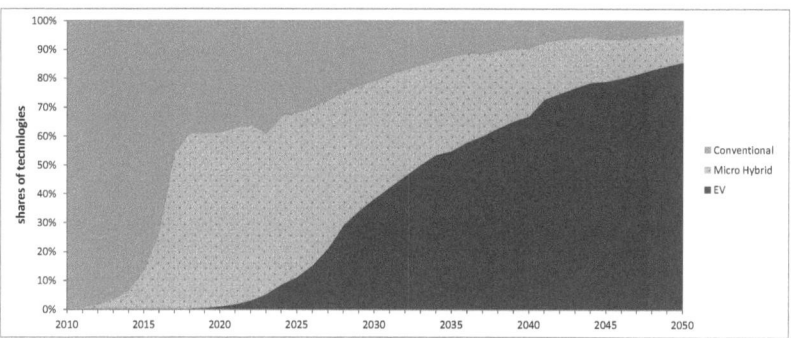

Figure B-18: Diffusion of propulsion technologies in the compact class - *BAU Scenario*

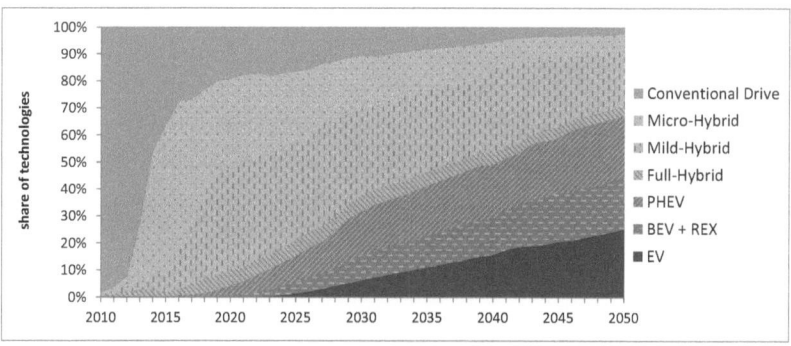

Figure B-19: Diffusion of propulsion technologies in the middle class - *BAU Scenario*

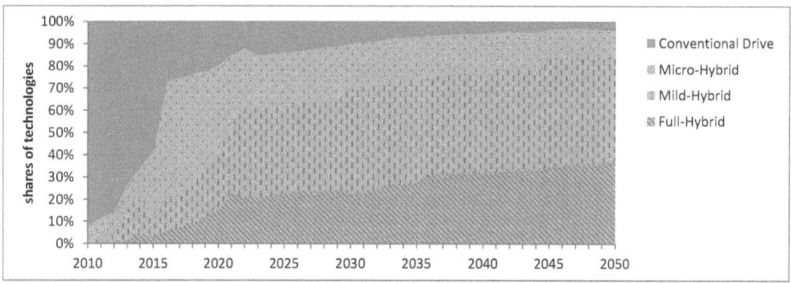

Figure B-20: Diffusion of propulsion technologies in the upper class - *BAU Scenario*

12.3.2 "Active" Policy & moderate fossil fuel price increase (Policy & Low Price-Scenario)

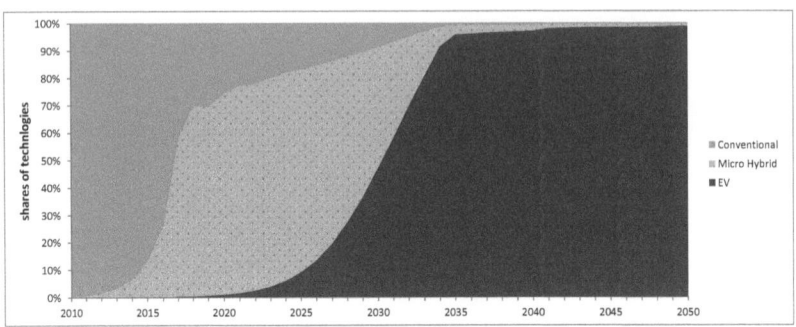

Figure B-21: Diffusion of propulsion technologies in the compact class - *Policy & Low Price - Scenario*

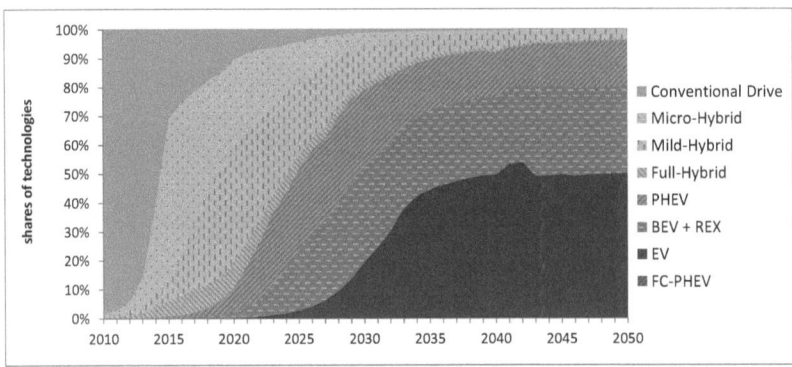

Figure B-22: Diffusion of propulsion technologies in the middle class – *Policy & Low Price – Scenario*

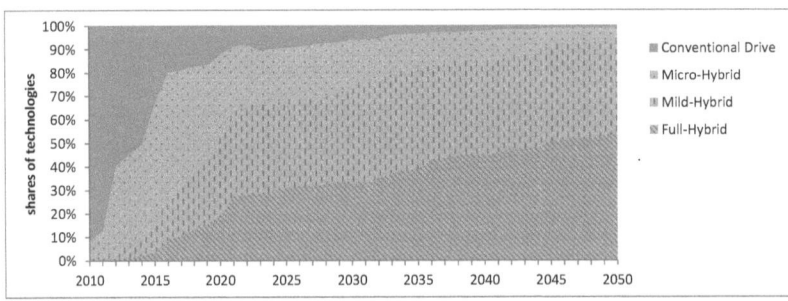

Figure B-23: Diffusion of propulsion technologies in the upper class – *Policy & Low Price - Scenario*

12.3.3 Fuel tax only & low price – Scenario

Figure B-24 shows the fleet development in Austria if only the fuel tax scheme of the Policy scenario was implemented. It shows that the fuel tax alone is a strong driver for the diffusion of electric propulsion technologies and it has almost the same effect on the long term fleet development as the combined measures in the Policy scenario.

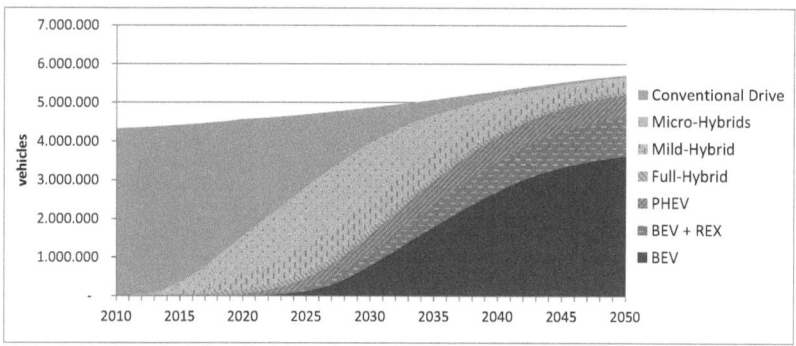

Figure B-24: Passenger car fleet – *"only fuel tax"* & Low Price – Scenario

12.3.4 Tax on acquisition only & low price – Scenario

Figure B-25 shows the development of the passenger car fleet if only the tax on acquisition of the policy scenario is implemented. Even though these measures have some mid- to long term impact on the diffusion of electrified propulsion technologies the effect is much weaker than it was for the fuel taxation measures. Also the effect on the long term fleet growth is minor.

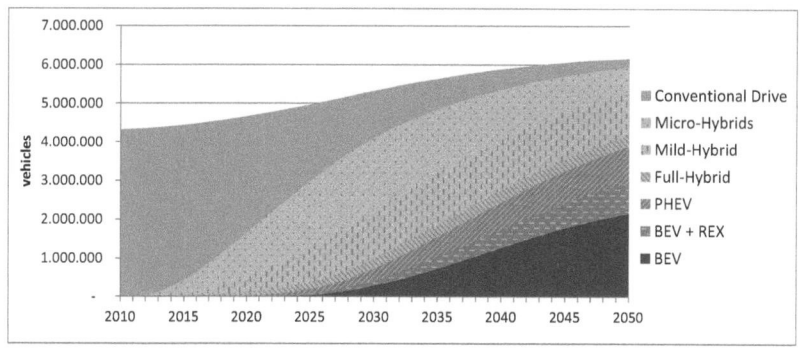

Figure B-25: Passenger car fleet – "only tax on acquisition" & Low Price – Scenario

12.3.5 "Business as usual"-Policy & substantial fossil fuel price increase (BAU & High Price Scenario)

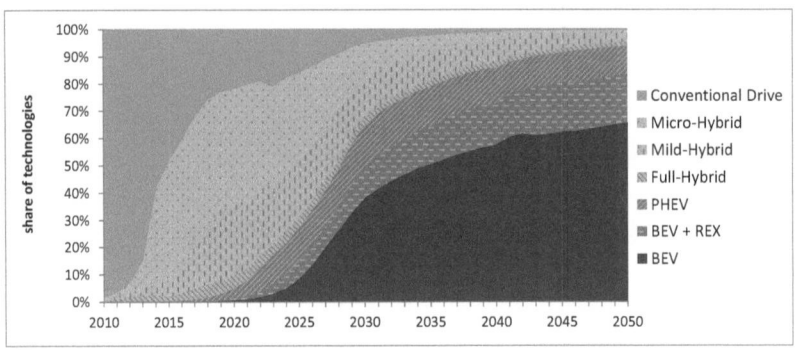

Figure B-26: Market diffusion of vehicle propulsion technologies BAU & High Price – Scenario

12.3.6 "Active" Policy & substantial fossil fuel price increase (Policy & High Price-Scenario)

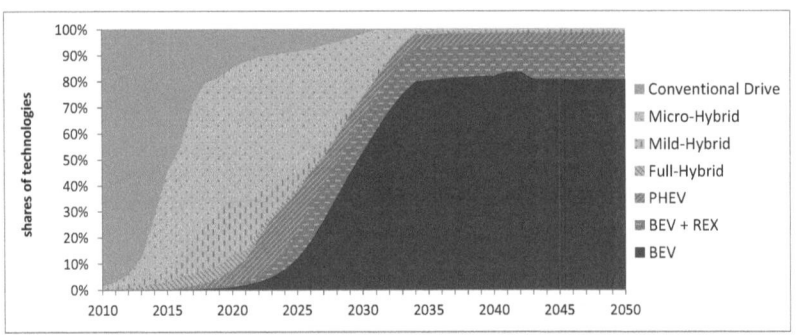

Figure B-27: Market diffusion of vehicle propulsion technologies Policy & High Price - Scenario

12.4 Shares of vehicle classes in the scenarios

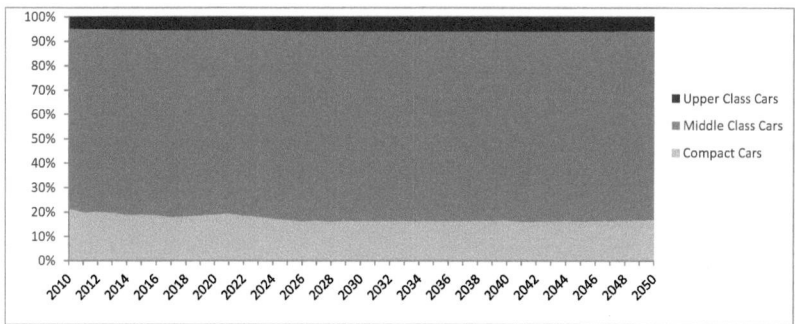

Figure B-28: shares of vehicle classes – BAU & Low Price - Scenario

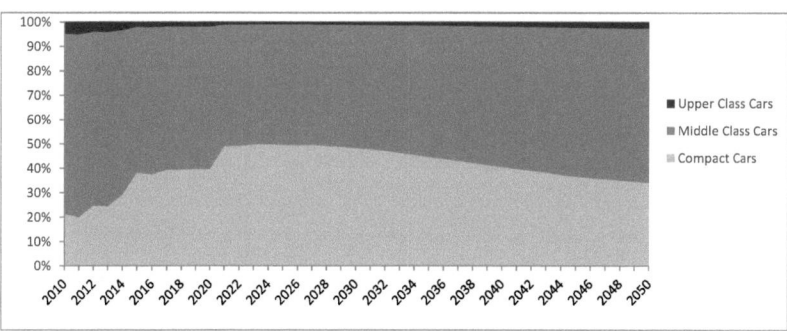

Figure B-29: shares of vehicle classes – Policy &Low Price - Scenario

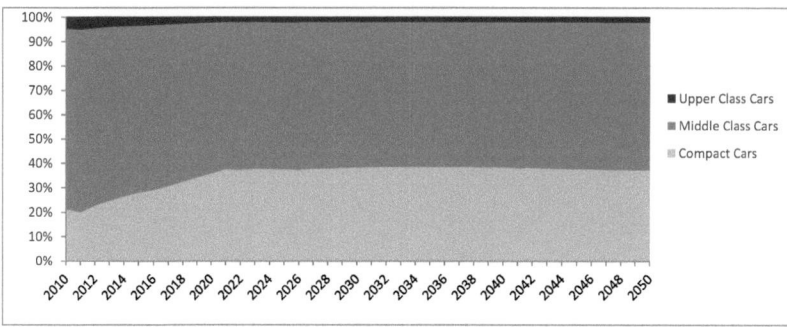

Figure B-30: shares of vehicle classes – BAU & High Price - Scenario

12.5 Fuel supply

Table B-3: shares of biofuel sources

			2010	2020	2030	2040	2050
Biofuel Blending							
	Share of Biofuels	BAU biofuel policy	5,75%	10%	10%	10%	10%
		Active biofuel policy	5,75%	10%	20%	25%	30%
Biofuel blends							
	Gasoline Blends	Ethanol 1	100%	75%	50%	25%	0%
		Ethanol 2	0%	25%	50%	75%	100%
	Diesel Blends	Biodiesel	100%	75%	50%	25%	0%
		BTL	0%	25%	50%	75%	100%
	CNG-Blends	Biogas	100%	95%	80%	80%	80%
		SNG	0%	5%	20%	20%	20%
Biomass Ressources for Biofuels							
	Ethanol 1	Corn	15%	15%	15%	15%	15%
		Wheat	75%	75%	75%	75%	75%
		Sugar Beet	10%	10%	10%	10%	10%
	Ethanol 2	Straw	100%	80%	60%	40%	40%
		Short Rotation Coppice	0%	20%	40%	60%	60%
	Biodiesel	Rapeseed	75%	75%	75%	75%	75%
		Sunflower	20%	20%	20%	20%	20%
		Used Cooking Oil	5%	5%	5%	5%	5%
	BTL	Wood Chips	100%	100%	100%	100%	100%
	Biogas	Manure	5%	4%	3%	2%	3%
		Energy Plants	35%	39%	43%	47%	50%
		Maize Silage	60%	57%	54%	51%	47%
	SNG	Wood Chips	100%	100%	100%	100%	100%

Figure B-31 to Figure B-34 show the fuel mix and the required biofuels in the BAU scenario with BAU biofuel blending and with active biofuel blending. To meet the Austrian biofuel demand of the active biofuel blending scenario biofuel production has to be increased by 300 % up to 2040.

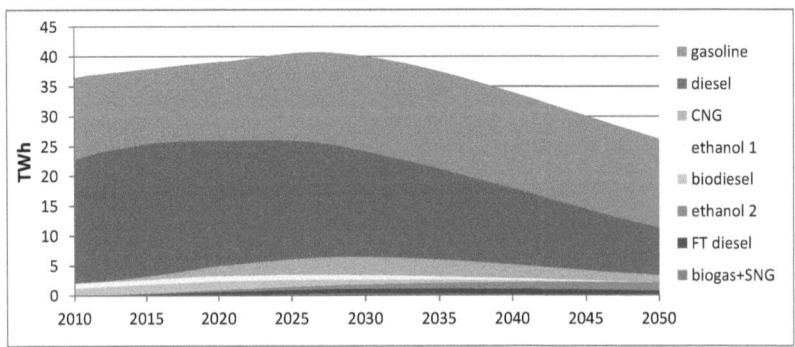

Figure B-31: Consumption of fuel types in the - *BAU Scenario & low fossil fuel price + BAU biofuel blending*

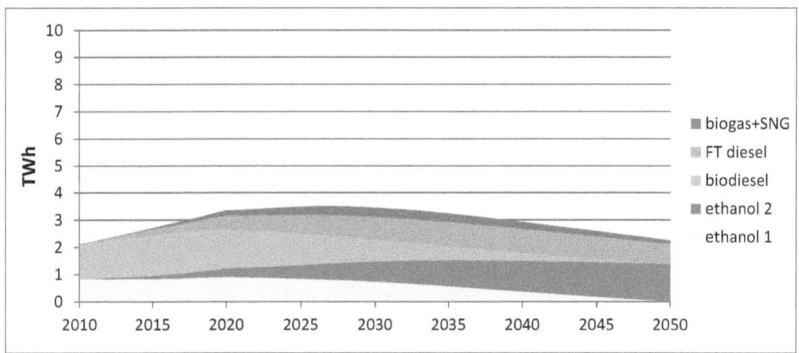

Figure B-32: Quantities of biofuels required & *BAU Scenario + low fossil fuel price + BAU biofuel blending*

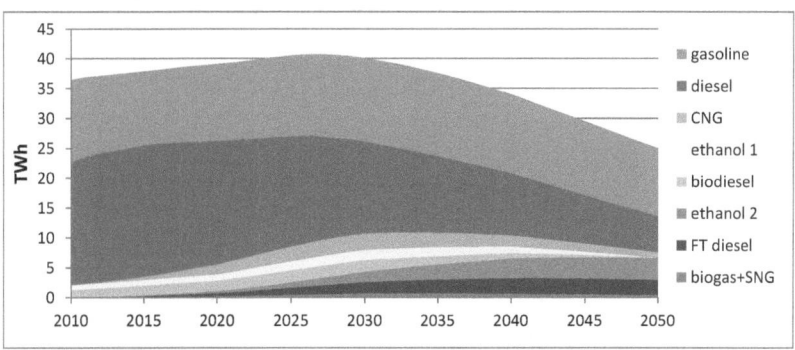

Figure B-33: Consumption of fuel types in the - *BAU Scenario & low fossil fuel price + Active biofuel blending*

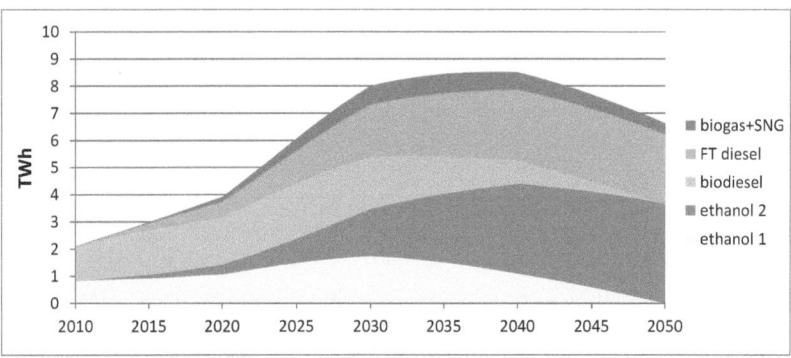

Figure B-34: Quantities of biofuels required + *BAU Scenario & low fossil fuel price + Active biofuel blending*

12.6 Electricity supply for EVs

Table B-4: data for load profile estimation 2020 & 2030

		yearly driving distance [km]	average work day daily driving distance [km]	average work day in electric mode	2020 energy consumption driving [kWh/100km]	2020 energy consumption plug [kWh/day]	2020 charging time [h/day]	2020 number of vehicles	2030 energy consumption driving [kWh/100km]	2030 energy consumption plug [kWh/day]	2030 charging time [h/day]	2030 number of vehicles	Plug-Power [kW]
Middle Class	PHEV-40	10,000	30	100%	21.3	7.1	1.9	674	20.4	6.8	1.9	54,181	3.68
Middle Class	BEV-65	10,000	30	100%	21.3	7.1	1.9	96	20.4	6.8	1.9	36,329	3.68
Middle Class	BEV-130	10,000	30	100%	21.3	7.1	1.9	-	20.4	6.8	1.9	24,220	3.68
Middle Class	PHEV-40	15,000	45	80%	21.3	8.5	2.3	21,586	20.4	8.2	2.2	247,151	3.68
Middle Class	BEV-65	15,000	45	100%	21.3	10.7	2.9	3,585	20.4	10.2	2.8	200,686	3.68
Middle Class	BEV-130	15,000	45	100%	21.3	10.7	2.9	518	20.4	10.2	2.8	118,837	3.68
Middle Class	PHEV-40	20,000	60	60%	21.3	8.5	2.3	22,906	20.4	8.2	2.2	137,302	3.68
Middle Class	BEV-65	20,000	60	100%	21.3	14.2	3.9	5,437	20.4	13.6	3.7	125,241	3.68
Middle Class	BEV-130	20,000	60	100%	21.3	14.2	3.9	999	20.4	13.6	3.7	101,438	3.68
Compact Class	BEV-50	10,000	30	100%	18.9	6.3	1.7	1,892	18.0	6.0	1.6	370,471	3.68
Compact Class	BEV-50	15,000	45	100%	18.9	9.4	2.6	4,357	18.0	9.0	2.4	299,554	3.68

Table B-5: data for load profile estimation 2040 & 2050

		yearly driving distance [km]	average work day daily driving distance [km]	average work day in electric mode	2040 energy consumption driving [kWh/100km]	2040 energy consumption plug [kWh/day]	2040 charging time [h/day]	2040 number of vehicles	2050 energy consumption driving [kWh/100km]	2050 energy consumption plug [kWh/day]	2050 charging time [h/day]	2050 number of vehicles	Plug-Power [kW]
Middle Class	PHEV-40	10,000	30	100%	19.6	6.5	1.8	93,696	18.7	6.2	1.7	85,590	3.68
Middle Class	BEV-65	10,000	30	100%	19.6	6.5	1.8	155,179	18.7	6.2	1.7	239,611	3.68
Middle Class	BEV-130	10,000	30	100%	19.6	6.5	1.8	256,544	18.7	6.2	1.7	462,322	3.68
Middle Class	PHEV-40	15,000	45	80%	19.6	7.8	2.1	234,842	18.7	7.5	2.0	180,335	3.68
Middle Class	BEV-65	15,000	45	100%	19.6	9.8	2.7	399,340	18.7	9.3	2.5	490,020	3.68
Middle Class	BEV-130	15,000	45	100%	19.6	9.8	2.7	631,942	18.7	9.3	2.5	993,915	3.68
Middle Class	PHEV-40	20,000	60	60%	19.6	7.8	2.1	113,646	18.7	7.5	2.0	91,904	3.68
Middle Class	BEV-65	20,000	60	100%	19.6	13.0	3.5	203,146	18.7	12.4	3.4	241,692	3.68
Middle Class	BEV-130	20,000	60	100%	19.6	13.0	3.5	347,458	18.7	12.4	3.4	504,097	3.68
Compact Class	BEV-50	10,000	30	100%	17.1	5.7	1.6	1,343,846	16.2	5.4	1.5	1,521,167	3.68
Compact Class	BEV-50	15,000	45	100%	17.1	8.6	2.3	651,472	16.2	8.1	2.2	656,323	3.68

Die VDM Verlagsservicegesellschaft sucht für wissenschaftliche Verlage abgeschlossene und herausragende

Dissertationen, Habilitationen, Diplomarbeiten, Master Theses, Magisterarbeiten usw.

für die kostenlose Publikation als Fachbuch.

Sie verfügen über eine Arbeit, die hohen inhaltlichen und formalen Ansprüchen genügt, und haben Interesse an einer honorarvergüteten Publikation?

Dann senden Sie bitte erste Informationen über sich und Ihre Arbeit per Email an *info@vdm-vsg.de*.

Sie erhalten kurzfristig unser Feedback!

VDM Verlagsservicegesellschaft mbH
Dudweiler Landstr. 99   Telefon +49 681 3720 174
D - 66123 Saarbrücken   Fax   +49 681 3720 1749
www.vdm-vsg.de

Die VDM Verlagsservicegesellschaft mbH vertritt

Printed by Books on Demand GmbH, Norderstedt / Germany